RESIDUE REVIEWS

VOLUME 31

RESIDUE REVIEWS

Residues of Pesticides and Other
Foreign Chemicals in Foods and Feeds

Editor
FRANCIS A. GUNTHER

Assistant Editor
JANE DAVIES GUNTHER

Riverside, California

VOLUME 31

Springer Science+Business Media, LLC
1970

ISBN 978-3-662-38889-1   ISBN 978-3-662-39818-0 (eBook)
DOI 10.1007/978-3-662-39818-0

© 1970 by Springer Science+Business Media New York
Originally published by Springer-Verlag New York Inc. in 1970.
Softcover reprint of the hardcover 1st edition 1970
Library of Congress Catalog Card Number 62–18595.

Title No. 7850

Preface

That residues of pesticide and other "foreign" chemicals in food-stuffs are of concern to everyone everywhere is amply attested by the reception accorded previous volumes of "Residue Reviews" and by the gratifying enthusiasm, sincerity, and efforts shown by all the individuals from whom manuscripts have been solicited. Despite much propaganda to the contrary, there can never be any serious question that pest-control chemicals and food-additive chemicals are essential to adequate food production, manufacture, marketing, and storage, yet without continuing surveillance and intelligent control some of those that persist in our foodstuffs could at times conceivably endanger the public health. Ensuring safety-in-use of these many chemicals is a dynamic challenge, for established ones are continually being displaced by newly developed ones more acceptable to food technologists, pharmacologists, toxicologists, and changing pest-control requirements in progressive food-producing economies.

These matters are of genuine concern to increasing numbers of governmental agencies and legislative bodies around the world, for some of these chemicals have resulted in a few mishaps from improper use. Adequate safety-in-use evaluations of any of these chemicals persisting into our foodstuffs are not simple matters, and they incorporate the considered judgments of many individuals highly trained in a variety of complex biological, chemical, food technological, medical, pharmacological, and toxicological disciplines.

It is hoped that "Residue Reviews" will continue to serve as an integrating factor both in focusing attention upon those many residue matters requiring further attention and in collating for variously trained readers present knowledge in specific important areas of residue and related endeavors; no other single publication attempts to serve these broad purposes. The contents of this and previous volumes of "Residue Reviews" illustrate these objectives. Since manuscripts are published in the order in which they are received in final form, it may seem that some important aspects of residue analytical chemistry, biochemistry, human and animal medicine, legislation, pharmacology, physiology, regulation, and toxicology are being neglected; to the contrary, these apparent omissions are recognized, and some pertinent manuscripts are in preparation. However, the field is so large and the interests in it are so varied that the editors and the Advisory Board earnestly solicit suggestions of topics and authors to help make this international book-series even more useful and informative.

"Residue Reviews" attempts to provide concise, critical reviews of timely advances, philosophy, and significant areas of accomplished or needed endeavor in the total field of residues of these chemicals in foods, in feeds, and in transformed food products. These reviews are either general or specific, but properly they may lie in the domains of analytical chemistry and its methodology, biochemistry, human and animal medicine, legislation, pharmacology, physiology, regulation, and toxicology; certain affairs in the realm of food technology concerned specifically with pesticide and other food-additive problems are also appropriate subject matter. The justification for the preparation of any review for this book-series is that it deals with some aspect of the many real problems arising from the presence of residues of "foreign" chemicals in foodstuffs. Thus, manuscripts may encompass those matters, in any country, which are involved in allowing pesticide and other plant-protecting chemicals to be used safely in producing, storing, and shipping crops. Added plant or animal pest-control chemicals or their metabolites that may persist into meat and other edible animal products (milk and milk products, eggs, etc.) are also residues and are within this scope. The so-called food additives (substances deliberately added to foods for flavor, odor, appearance, etc., as well as those inadvertently added during manufacture, packaging, distribution, storage, etc.) are also considered suitable review material.

Manuscripts are normally contributed by invitation, and may be in English, French, or German. Preliminary communication with the editors is necessary before volunteered reviews are submitted in manuscript form.

Department of Entomology   F.A.G.
University of California
Riverside, California
December 19, 1969

Table of Contents

Leaf structure as related to absorption of pesticides and other compounds*

By

HERBERT M. HULL * *

Contents

*See Table I for common or trade names and chemical nomenclature of the compounds mentioned, and Table II for a glossary of botanical terms.
**Crops Research Division, Agricultural Research Service, *U.S. Department of Agriculture* and *Department of Watershed Management*, Arizona Agricultural Experiment Station, Tucson, Arizona.

I. Introduction

Plants are sprayed and dusted with chemicals for numerous reasons. Some of the chemicals applied to plants exert their influence simply by remaining on the foliage surface; others must be absorbed and perhaps translocated before their activity is realized. Many of the compounds used have a mammalian toxicity less than that of table salt, whereas others exceed strychnine in toxicity. Likewise, many are relatively non-phytotoxic but others kill plants readily. Possible foliar residues which may be left by various pesticides are of course influenced by both absorption characteristics and metabolic breakdown. The various plant, environmental, and chemical factors which affect absorption are examined in this review. A variety of pesticides is considered, as well as certain growth regulators and other substances which are not, strictly speaking, "pesticides." A limited consideration is also given to recent findings relating to penetration of inorganic ions, where the mechanisms involved may bear certain similarities to the penetration of some organic substances which also ionize or hydrolyze during the course of their penetration. Because of this broad coverage, space precludes detailed discussions of all aspects of absorption and of associated structural relationships. These subjects are developed quite fully in many of the reviews listed below and in the introductions to some of the specific sections in this report. Also, only foliar absorption and movement within the leaf mesophyll and veins are considered. Long distance transport in the vascular system of the stems and roots is not included, since several excellent reviews and monographs on this subject have recently been published or are in preparation.

Various aspects of foliar absorption of pesticides and miscellaneous organic substances have been reviewed by CRAFTS and FOY (1962), EBELING (1963), FRANKE (1967 a), HOSKINS (1962), HULL (1964 a), LINSKENS et al. (1965), and MARTIN (1966). Coverage more specifically relating to herbicides includes that of CRAFTS (1964), CURRIER and DYBING (1959), DAY et al. (1968), FOY (1964 a), FOY et al. (1967), HAMMERTON (1967), and SARGENT (1966). Herbicidal selectivity as a function of absorption has been discussed by L. A. NORRIS (1967); the absorption characteristics of numerous individual herbicides are also available (HULL et al. 1967).

Reviews describing the more general aspects of absorption and translocation of plant growth regulators includes the work of AUDUS (1967), MITCHELL et al. (1960), MITCHELL and LINDER (1963), VAN OVERBEEK (1956), and SARGENT (1965). Of additional interest to researchers working with these compounds is the recent publication of a revised manual detailing techniques for studying the absorption and transport of hormones and regulators (MITCHELL and LIVINGSTON 1968).

The foliar absorption of insecticides is less well documented, although some information on this subject is available in the work of FINLAYSON and MAC CARTHY (1965) and that of D. M. NORRIS (1967); the latter author cites reviews covering foliar uptake of systemic insecticides. The uptake of chemotherapeutants, including antibiotics and various fungicides and bactericides, has been covered by P. W. BRIAN (1967), DIMOND (1965), GOODMAN (1962), and WAIN and CARTER (1967). Advances in the foliar absorption of mineral nutrients has been thoroughly documented by JYUNG and WITTWER (1965), WITTWER (1964 and 1968), WITTWER et al. (1967), and WITTWER and TEUBNER (1959). Literature relating to both foliar and root absorption of radionuclides has been reviewed by RUSSELL (1963).

Although information in the above reviews emphasizes absorption of the substances indicated, secondary emphasis varies greatly. Some consider detailed leaf or cuticle structure; others emphasize surface structure and wetting characteristics. Some include more information on translocation and metabolism than on absorption. Many consider only research on intact plants, whereas others describe work with isolated leaves, leaf slices, stem segments, isolated cuticle, and even isolated cells. The present review will deal primarily with intact plants or isolated leaves, although the other techniques are touched on lightly in the appropriate sections, wherever it is felt that they will contribute to the overall picture.

In this discussion specific classes of pesticides or other substances are not considered separately, since there can be as much variation in absorptive charactreistics of individual substances within a single pesticide classification (e.g., insecticides) as exists among individuals of different classes. Research on pesticide absorption has accelerated greatly in recent years, as judged by the extensive publication record. Work on growth regulators and herbicides, in particular, seems to have mushroomed. Usage figures tend to parallel the research picture. For example, in West Germany, organic herbicides accounted for 38 percent of the total quantity of pesticides used during 1966, as compared with only 5.3 percent in 1952 (DREES 1967). The same trend has undoubtedly occurred in most other countries. In 1967, the value of herbicide sales in the United States exceeded the value of insecticide sales for the first time, having increased nearly 67 percent over the 1966 value (MAHAN et al. 1968). Because of the great increase in literature, it was not possible to include even half of the publications that would have been pertinent to this review. If the author has overlooked any significant papers, particularly monographs or reviews, he offers apology and would appreciate notification of the omission.

Structure of the various foliar constituents will be considered in the order in which they would be encountered by a penertating molecule. Thus the surface wax is contacted first, followed by the remainder of the cuticle, the epidermal cell wall and possibly the protoplast, and finally the leaf mesophyll and leaf veins. For information on long-distance translocation within the conductive tissues of stems and roots other sources of information must be consulted.

II. The cuticle

a) *Epicuticular wax*

1. **Structural and chemical characteristics; biosynthesis.**—The fine structure of surface wax has received comprehensive review by EGLINTON and HAMILTON (1967). They discuss possible origins of surface

wax crystals, as well as the mechanism whereby they may be extruded. They also describe De Bary's early classification of waxy coatings into four types, based on his studies with the light microscope. More recently, AMELUNXEN et al. (1967), after studies with the stereoscan electron microscope, have suggested six classifications for surface wax. These include: 1) a simple granular wax crust, 2) wax rods and filaments, 3) wax plates and scales, 4) wax layers and crusts, 5) aggregate wax coatings, and 6) liquid or viscous wax coatings. The excellent depth-of-field obtainable with the scanning electron microscope (even though its resolution does not approach that of the regular instrument) makes it an ideal tool for the study of surface features, as clearly demonstrated by CHAPMAN (1967), ECHLIN (1968), and HESLOP-HARRISON and HESLOP-HARRISON (1969) in their micrographs of leaves, seeds, pollen, etc. The scanning electron-micrographs of isolated cuticles just published by LANG (1969) clearly depict microrelief on both outer and inner surfaces, the latter showing various degrees of ribbing at the junction of the epidermal cells.

Surface wax on both leaves and fruits of *Vitis vinifera* was observed by POSSINGHAM et al. (1967) to be morphologically similar, consisting of a series of overlapping platelets. Exposure to light petroleum vapor for 30 seconds disorganized the platelet structure and markedly increased subsequent cuticular transpiration. BAIN and McBEAN (1967), studying *Prunus domestica* fruits, observed that the surface wax occurred as a two-layered structure. The inner layer consisted of a matrix of thin platelets, while the outer layer was composed of fragile projections—many of which appeared tubular. Although the wax deposits remained uniform at about 300 μg./cm.^{2} during development, the incidence and complexity of the outer projections continuously increased. Studies of this nature, including techniques of altering the surface wax configuration, are important in minimizing drying time in the preparation of dried fruits. It is of interest that when the surface wax deposits of apple fruits are completely removed with a suitable solvent, the underlying cuticle shows a structure apparently related to the shape of the wax above it, as demonstrated by the electronmicrographs of HALL (1966). Thus, Sturmer Pippins showed circular shaped ridges, Golden Delicious had channels which existed under the lath-like wax deposits, while other varieties exhibited no distinctive features.

Surface wax deposits and intercalated wax layers in the outer portion of the cuticle are difficult to demonstrate in cross section by electron microscopy, both because of their tendency to dissolve in the usual embedding media and their instability in the electron beam. However, SITTE (1962) has demonstrated fine lamellations in the leaf cuticle of *Ficus elastica*, the alternating layers apparently consisting of weakly lipophilic cutin and water repellent wax. The wax lamellae were generally only about three nm. thick, corresponding to a monomolecular film having the rod-shaped molecules situated at right angles to the lamellar plane. This structure was also suggested by the optical properties under polarized light. AMBRONN (1888) was perhaps the first to note negative birefringence in the outer portion of the cuticle, apparently due to the orientation of the wax molecules, and to further observe that the birefringence (and orientation) was lost when the specimen was heated to 100°C. Cuticular lamellations have also been noted in *Eucalyptus* sp. leaf transections by HALLAM (1964), and were observed to decrease in thickness toward the external surface. Three species of this tree were investigated both by cross sections and surface replication, and proved to have either "tube" or "plate" type surface wax structures, or a mixture thereof. By evaporation of gold-palladium directly on the leaf surface before dehydration and embedding, it was possible to preserve surface structure in the cross sections, which could then be compared with the carbon surface replicas. In this species wax may migrate to the leaf surface through anastomosing channels between the cuticular lamellae. Additional investigations on the electron microscopy of the waxes of cabbage leaves, and the manner in which electromagnetic radiation produces a tempering of this constituent have been reported by HANSZEN (1961).

In his studies of *Agave americana* cuticle, CRISP (1965) found that the use of a cold stage in the electron microscope was most helpful. The high vacuum and temperatures of 200° to 250°C. produced when the beam is at saturation cause a melting, evaporation, or sublimation of the wax. This loss may be accompanied by microexplosions, resulting in destruction of the tissue. By maintaining the specimen below −180°C., the cold stage greatly minimizes this problem. It does, however, act as a cold trap and thus increases the rate of contamination on the specimen to some extent. The difficulties involved in adequate preservation and identification of free lipids by electron microscopy have been further outlined by CASLEY-SMITH (1967), who notes that only triglycerides fixed by postchromation and lipofuscins are able to survive extraction during the embedding process. Also, after glutaraldehyde fixation, the tissue may be bathed in certain combinations of mutually reacting salts which are more soluble in lipids than in water, and the reaction of which yields a precipitate insoluble in both lipid and aqueous solvents. Some combinations were fairly specific for neutral lipid and did not stain other regions of the cell. Several new embedding media, including a urea-formaldehyde compound, were found to preserve small lipid droplets, but were still not completely satisfactory.

The chemistry of the epicuticular waxes has been discussed in detail by EGLINTON and HAMILTON (1967), and KREGER (1958), as has the general biosynthesis of surface lipids (KOLATTUKUDY 1968 a) and more specifically the fatty acids (WOODBINE 1964). Therefore, the subject will be only very briefly summarized here, and a few additional recent findings considered. Hydrocarbons of the surface lipids are for the most part *n*-paraffins, which may contain from 21 to 35 carbon atoms. Although it was earlier considered that this fraction consisted exclusively of odd-numbered homologues, mass spectrometry has demonstrated that in some cases even-numbered members may be present as minor components (WALDRON et al. 1967). A single alkane often predominates, but several may be present. For example, LASETER et al. (1968 a) demonstrated that the *n*-alkanes of cabbage leaf lipids ranged from C_{27} to C_{31}, with C_{29} comprising the major component (92 percent) of the *n*-heptane fraction. Additional work (LASETER et al. 1968 b) established that the hydrocarbons making up the surface wax of *Ustilago maydis* (Basidiomycetes) consisted predominantly of C_{25}, C_{27}, and C_{29} *n*-alkanes. After studying the uptake of numerous ^{14}C-labeled compounds by *Wistaria sinensis* flowers, and their incorporation into the C_{27} paraffin, heptacosane, BRECCIA and ABBONDANZA (1967) concluded that the chain was lengthened by condensation of acetate, as well as by condensation of larger units such as nonanoate. Mevalonate, however, was utilized only after being broken down into simpler molecules; and pyruvate and succinate were utilized in pathways outside the Krebs cycle. Other compounds participated only indirectly in heptacosane synthesis, possibly by the donation of ^{14}CO$_2$ following their decarboxylation. KOLATTUKUDY (1968 b) has presented evidence suggesting that elongation of a common fatty acid such as palmitic to very long-chain fatty acids of appropriate length, followed by decarboxylation, appears to be the most likely pathway for paraffin biosynthesis. Enzymes involved in such a synthesis, however, have not been isolated. Although fatty acid synthesis is not affected by TCA, even very low concentrations ($10^{-5}M$) strongly inhibit paraffin synthesis. Fatty acid synthesis was observed to occur in the mesophyll and to be tightly coupled to photosynthetic reactions, whereas the elongation process responsible for synthesis of paraffins and very long fatty acids occurred in the epidermis, and was apparently not so coupled. The fact that virtually all leaf paraffins are found in the cuticle, where they could have been synthesized within the epidermis (which usually contains no chloroplasts), would seem compatible with this picture.

The free fatty acids and alcohols of the surface lipids as well as their esters generally have an even number of carbon atoms with a chain length ranging from C_{10} to C_{30}. Odd-numbered members do exist in some lipids, however, as pointed out by WALDRON et al. (1961) in the case of primary alcohols. Also of interest, is the report of this group that of the numerous species investigated, the only isoparaffins found were either in rose petal or tobacco leaf wax. Isoparaffins have also been detected in the surface wax of tobacco leaf by MOKHNACHEV et al. (1967), although in considerably lesser quantity than normal paraffins. They found paraffins containing from 14 to 34 carbon atoms, with C_{27}

and C_{31} prevailing, and noted that those with an odd number of carbon atoms were twice as plentiful as those with an even number. KOLATTU-KUDY (1968 c) recently observed that in plants capable of synthesizing branched paraffins, such as tobacco, the relative amounts of the substances synthesized may depend on the availability of branched precursors—which in turn is controlled by the metabolic balance of the branched amino acids. Other species are largely incapable of such synthesis. *Brassica oleracea* leaves, when fed ^{14}C-labeled isobutyrate and isoleucine through the petioles, were not capable of incorporating more than one-third of the ^{14}C into C_{29} paraffin and derivatives of the surface lipids, whereas over two-thirds of the ^{14}C from straight chain precursors was usually incorporated. The small amount of activity occurring in the paraffin fraction was found in the n-C_{29} fraction rather than branched paraffins, suggesting that the ^{14}C in the paraffin must have come from degradation products.

A great variety of saturated and unsaturated fatty acids have been found in the surface lipids. Thus, LASETER *et al.* (1968 a) have identified linolenic, linoleic, and palmitic acids as the major constituents of cabbage leaves, making up over 50 percent of the total fatty acids present. The cuticle oil from ripe apples is made up largely of fatty acids. DAVENPORT (1960) found such oil to contain 48.5 percent fatty acids. This fraction consisted predominantly of oleic, linoleic, and linolenic acids, with minor amounts of palmitic, myristic, lauric, caprylic, enanthic, and caproic acids. Earlier work had demonstrated that this oil fraction increased three to four fold during storage of the apples, and was accompanied by an increase in the resistance of the skin to gaseous diffusion. The extreme complexity of the above-described chemical mixtures, as pointed out by EGLINTON and HAMILTON (1967), must surely influence the structure of the surface deposits appearing on the different organs of numerous plant species. This structure markedly influences wettability characteristics, as will be discussed later.

2. Ontogenetic changes in wax and the effect of environment upon its development; taxonomic aspects.—It is known that certain chemical changes occur in the constituents of the surface wax during foliar development, as well as in the total leaf lipids. Differences which exist among certain species, however, preclude the drawing of an overall picture. MATSUDA (1962), in sudying the biosynthesis of wax from the candelilla plant (*Euphorbia antisyphilitica*), observed that waxes from young tissues had a lower percentage of total paraffins but higher percentages of acids and alcohols than did waxes from older tissues. The relative percentages of the C_{29} to C_{34} homologues making up the paraffin fraction, however, remained relatively stable regardless of tissue age. Such stability would be supported by the finding of KURTZ (1950) that wax melting points of some 13 plant species did not change markedly with age, although some species showed a slight increase in melting point during maturity which was correlated with a decrease in unsaturation.

Such a decrease was apparently due to oxidation by atmospheric oxygen. He also noted that the wax acids decreased rapidly in young plants but that both esters and acids increased with age, in contrast to the trend of acids in the candelilla plant previously described. Such increases may have been related to decreases in the nonwax acids sometimes observed during maturation. RADUNZ (1966), studying developmental relationships of fatty acids in *Antirrhinum majus*, observed a seven percent drop in total fatty acids during maturation. However, he also found, as did KURTZ, an increase in the proportion of saturated fatty acids during development—from 13 to a final 32 percent of the total fatty acids in the case of this species. In addition, he observed the strange relationship of the C_{18} acids decreasing with leaf development, while those with 12, 14, 15, 16, 17, 20, 22, 23, and 24 carbons increased. The major C_{16} and C_{18} fatty acids in tobacco were found by CHU and Tso (1968) to reach a maximum in the upper young leaves about 75 days after transplanting, whereas in older leaves the acids continuously declined, as they were observed to do in the previously described candelilla plant. These investigators also observed a significant increase in the highly unsaturated linolenic acid with leaf development, accompanied by decreases of the more saturated acids. Such progressive overall desaturation is in contrast to the increased saturation described by KURTZ (1950) for some other plant species. During studies on the leaf lipids of *Myosotis scorpioides*, a marsh plant, JAMIESON and REID (1968) noted relatively large proportions of γ-linolenic and octadeca-tetraenoic acids, as well as the more usual α-linolenic, linoleic, and palmitic acids. Changes in the proportions of all the different fatty acids were found during the growing season, and differences were also observed depending upon growth locality.

Environmental factors apparently influence the nature of surface wax in several ways. MARTIN (1961) reported no significant differences in content or composition of leaf wax from apple rootstocks grown in the greenhouse as compared with that from plants grown in the open, although the leaves produced under glass were 30 percent larger. Preliminary observations suggested that surface deposits per square cm. of leaf surface were characteristic of the variety, and were relatively unaffected by environment. An environment inducing slower growth was found by JUNIPER (1960 a) to create leaves having a thickened layer of surface wax projections, which in turn may well contribute to the observed increased resistance to both mechanical injury and herbicidal treatment. He also noted a correlation between wax production on pea leaves and light intensities over the range of 900 to 5,000 footcandles, the higher intensities giving a white waxy bloom. Wettability (as determined by contact angle and discussed in a subsequent section) was, however, unaffected by the different degrees of bloom development. It was concluded that variation in light intensities encountered in the field would be unlikely to affect surface wax structure to the extent that it might influence wettability, although altered resistance to weathering might well be affected.

Variation in leaf surface waxiness between field populations of blue tussock grass (*Poa colensoi*) growing in the South Island of New Zealand has been reported by DALY (1964). When seeds from these populations were germinated and grown together in the greenhouse with ample soil moisture, the quantity of surface wax produced per gram fresh weight was highly and negatively correlated with both annual and summer rainfall and positively correlated to some extent with temperatures which existed at the field sites of collection. Electron microscopy of the leaves demonstrated that the glaucous surfaces of plants from semiarid habitats had rod- or spicule-like crystalline structures, whereas the greener leaves of plants growing in regions of greater rainfall possessed wax which was in the form of rather sparse non-crystalline smears.

There appears to be evidence of an interrelationship between climate of habitat or immediate day and night temperatures under which a plant is grown, and both the quantity and chemical composition of the surface wax. McNAIR (1931) reported that the constituents of the wax from tropical plants generally had a greater molecular weight than those from plants of more temperate regions. The higher melting point of foliar wax from *Prosopis juliflora* plants grown at relatively high day and night temperatures, as compared to those grown at lower temperatures, is also suggestive of a positive correlation between temperature and average molecular weight of the wax constituents (HULL 1958). A direct relationship of percentage of total wax with temperature was reported in this work, as it also was by SKOSS (1955) in the case of tobacco. He observed that although maximum cuticular development occurred at median day temperatures of 17°C., greatest percentage of wax was produced at the higher 23° to 30°C. temperatures. He also noted that both wax and cutin components of the cuticle were approximately 20 percent greater (weight/unit area) in *Hedera helix* growing outdoors in the sun as compared to the shade.

Relatively little information appears to be available relating to structural damage of leaves and fruits as a result of phytotoxic pesticidal sprays. LINSKENS *et al.* (1965) have discussed this aspect, including work indicating that susceptibility to phenyl mercuric acetate is related to differences in cutin rather than to differences in wax deposition, or the presence of a thin membrane below the surface wax. Certain pesticides and other substances do, however, appear to have the ability to alter surface wax structure to some extent, according to WORTMANN (1965). Marked changes in structure on the surfaces of *Brassica napus* and wheat leaves occurred following spraying with parathion, MCPA, or a wetting agent. The degree of change was dependent upon concentration, and was associated with an increased wettability. An ester and salt of MCPA had different effects on the structures, wettability being greater following the ester treatment than after the salt treatment. When treated only with the surfactant at 0.02 percent, the structures (particularly the rods) were largely destroyed, but regenerated in an entirely new multi-branched form. Depending upon the plant species, and the pesticide used and its concentration, the wax structures may or may not regenerate. Changes of this nature are no doubt responsible for a decrease in diffuse reflection of the leaf surface and a simultaneous increase in reflected gloss, as found by LINSKENS et al. (1965) to be associated with pesticide formulation.

It has been known for some time that air pollution produces certain anatomical anomalies in plant foliage. From the work of BYSTROM et al. (1968), it now appears that a specific action on the surface wax may also occur. On the surfaces of very young Beta vulgaris leaves, they observed wax generally to be in the form of stubby rodlets. With cell enlargement, rod extrusion continued, and the rods would coalesce to form plaques. In young-mature cells, a central plaque was surrounded by peripheral rodlets. After exposure of plants to air containing solar-irradiated automobile exhaust (smog), morphological changes occurred in the leaf surface wax which were apparently related to changes in the rate of wax rodlet extrusion. They also noted an anomalous surface wax pattern following an attack by aphids. The excessive waxy accumulations caused by such injury formed more slowly than did those resulting from smog damage. The circular accumulations were about a half μ in diameter, and surrounded the punctures made by the aphid proboscises.

In addition to the effect of foliar sprays and air pollutants on the structure of surface wax, certain compounds applied usually to the soil have been found to inhibit surface wax production. The mechanism of action of such compounds, including dalapon, dinoseb, monuron, and TCA, has been discussed by KOLATTUKUDY (1968) and by several other of the recent reviews mentioned in the introduction. GENTNER (1966) has also shown that EPTC and certain other thiolcarbamates can inhibit deposition of foliage wax on Brassica oleracea leaves. The inhibited deposition resulted in increased transpiration, increased spray retention and decreased contact angle of spray droplets, as well as enhanced absorption of foliar applied dinoseb. If EPTC was originally applied as a spray, only leaves then in the bud were affected, whereas granules applied to the soil extended the period during which formation of foliar wax was inhibited.

The relatively recent use of the surface wax constituents as a chemotaxonomic tool should perhaps be mentioned. This subject has been thoroughly discussed by EGLINTON and HAMILTON (1967). Their colleagues (PURDY and TRUTER 1961) had previously demonstrated by means of thin-layer chromatography of the ether extract of leaves of

numerous plant species, that patterns existed which were characteristic
of the individual species. These patterns generally did not change significantly with leaf age, and R_f values of the numerous components
were similar when the wax was extracted from different parts of the
plant. In a series of recently published papers, HERBIN and ROBINS
(1968 a, b, and c) and DYSON and HERBIN (1968) have greatly advanced our knowledge in this field. They have studied variations in the
alkanes, and in some cases the alkenes and ω-hydroxy acids of numerous
species in the Agavaceae, Crassulaceae, Cupressaceae, Liliaceae, Myrtaceae, and Pinaceae families. In the genus *Aloe*, for example, the alkanes of 63 species were investigated with gas chromatography. Species
specificity in composition was confirmed, and a correlation between
composition and sub-classification of the genus according to a preexisting standard was discernible in the leaf waxes and especially in
the perianth wax alkanes.

3. **Distribution on and within the leaf.**—Although it is generally
considered that foliar waxes are almost entirely incorporated in the
cuticle, this is not true for all species. In an investigation of various
Argentinian palms, DE GUTH (1966) determined that the leaf waxes of
Copernicia alba were totally epidermal, whereas in three *Trithrinax*
species, 34 to 54 percent of the waxes were in the mesophyll. She
also observed that certain wax characteristics depended upon origin,
e.g., epidermal waxes had a higher melting point and a lower acid
number than did those in the mesophyll. In their anatomical studies
of the carnauba palm (*Copernicia prunifera*), ARRAES *et al.* (1966)
found that the wax coating commonly reached 20 μ on both leaf
surfaces, and apparently increased to about 32 μ during drying of the
young leaves. This uniform and continuous coating was not interrupted
over stomata, although its texture was somewhat looser over such
areas.

On leaves which contain crystalline surface structures, the distribution of these structures is generally quite uniform (HOLLY 1964). However, variation may occur in the vicinity of stomata, where the wax is
often more sparsely distributed. Thus, WORTMANN (1965) noticed that
the rodlet and plate wax structures found on the surface of *Brassica
napus* leaves were almost absent around the stomata. On *Triticum aestivum* leaves, structureless areas appeared opposite the longitudinal
axes of the stomata, and only scattered structures were present around
the stomata of *Beta vulgaris* leaves. Structures were not discernible on
the cotyledon surfaces of any of these three species. JUNIPER (1960 b)
likewise observed that the density of surface wax platelets in the vicinity of stomata of *Chamaenerion angustifolium* was considerably less
than the density further away from the stomata. However, this type of
distribution is not universal. Electron micrographs of the lower surface
of *Prosopis juliflora* leaflets clearly demonstrate a density of wax struc-

tures on the guard cells approximately equivalent to that on adjacent epidermal cells (HULL 1964 a). The structures appear to even extend partially into the stomatal pores. Tubular waxy outgrowths on the surface of *Pinus radiata* needles are apparently distributed quite evenly (LEYTON and ARMITAGE 1968), except on the inner surface normally enclosed by the sheath. This inner enclosed surface was found to be the least wettable portion of the needles, apparently due to differences in chemical composition of the wax or to the absence of weathering. Premature exposure of young needles by removel of the basal sheath caused an outgrowth of the tubular waxy structures. When the needles were immersed in water, very little was taken up into the basal region normally enclosed by the sheath, as compared with the exposed surfaces. This absorption pattern is in contrast to the previously reported finding (LEYTON and JUNIPER 1963) that water uptake through the cuticular surface of *Pinus sylvestris* needles was more than three times greater below the sheath, where only surface wax platelets were present, than it was on the exposed portion which had tubelike waxy outgrowths. It is of interest that wax deposition apparently reaches a maximum during the second year of needle growth, as observed by FISHER *et al.* (1969) in their ultrastructural investigation of cuticular development in nine coniferous species. The deposited wax is subsequently sluffed off and additional deposition ceases.

Vertical distribution of the outer layered wax, often intercalated with the cutin and perhaps the deeper cuticular constituents in some cases, has been established by polarized light microscopy and X-ray interference, as comprehensively described by FREY-WYSSLING (1953) and some of the other reviews mentioned. In their recent investigation of *Pyrus communis* leaf cuticle, NORRIS and BUKOVAC (1968) determined that the upper cuticular membrane contained an almost continuous layer of negatively birefringent wax, this layer being discontinuous in the lower cuticle. The wax could be removed only by prolonged (18 hour) extraction with chloroform. In the upper cuticle it showed some degree of layering, with intervening bands of isotropic materials. Occasional wax pockets were deeply embedded in the lower cuticle. These embedded birefringent waxes decreased to varying extents over the anticlinal wall areas, and appeared broken in the upper cuticle wherever it overlay major veins. If the decreased birefringence in such areas represents a reduced quantity of embedded waxes rather than simply an altered molecular orientation, the apparent areas of lesser wax correlate well with observations showing increased penetration in such areas—even though the total cuticular thickness may be greater above veins than above mesophyll tissue.

4. **Effect on wettability, absorption, and water loss.**—There is little question but that both the chemical nature and physical structure of surface wax markedly influence the contact angle of impinging droplets

of spray solutions and consequent wetting.[1] SILVA FERNANDES (1965 a),
studying both surface ultrastructure and water-repellency of apple,
banana, black currant, broad bean, cabbage, coffee, hydrangea, pea,
rose, *Eucalyptus, Exochordia,* and *Rhus* leaves, was able to classify
them into groups which were water-repellent, easily wetted, or those
over which water spread without wetting. The affinity of the surfaces
for water was influenced both by amount of wax and its chemical composition, and especially by its physical configuration on the surface.
Waxes with significant quantities of long-chain ketones and paraffins
were the most difficult to wet, regardless of the quantity of wax present.
SILVA FERNANDES *et al.* (1963) had previously shown that the surface
wax of apple leaves, which consisted predominantly of esters and ursolic
acid, existed in greater quantity/unit area in young leaves than in fully
expanded leaves. Of the individual wax constituents which were isolated, paraffins and ursolic acid were most water-repellent. Ease of wetting the leaf surface was inversely correlated with the quantity of these
components/unit leaf area. However, even more important in influencing wetting was the amount of paraffin actually present on the wax
layer and its configuration. TROUGHTON and HALL (1967) observed extensive wax deposits on wheat, irrespective of variety, growth stage, or
part of the plant. These were in the form of platelets on both leaf surfaces of seedlings and some mature plants, whereas wax rods covered
the ear, culm, sheath, the lower surface of the flag leaf, and, occasion-
ally, the mature vegetative leaves. Contact angles on the upper leaf surfaces were higher than on the lower, except in the case of greenhouse-
grown plants, where there was no difference between the two sides.
Seedlings had higher contact angles than mature plants, but there was
no relationship of contact angle with tissue age within a leaf or within
a mature plant. The contact angle on the flag leaf of one cultivar was
24° higher on a greenhouse plant than it was on similar plants in the
field. These findings suggest that surface wax is well developed in very
young tissue and in greenhouse-grown plants. This is in contrast to
development of the entire cuticle, as will be discussed later.

An additional influence exerted by the microstructure of the leaf
surface is its ability to fix radioactive materials. MERTEN and BUCHHEIM
(1967) have established that hydrophobic surfaces have an uptake
capacity for such substances which is smaller by about one order of
magnitude than that of hydrophilic surfaces of the same type. In studying retention of foliar-applied [85]Sr, AMBLER and MENZEL (1966) likewise observed that retention differences among nine different species
were related to differences in wettability. Absorption and retention of
this radioisotope was also directly related to the relative humidity and
temperature of the air to which the leaves were exposed for 24 hours
after the applied solution became dry.

---

[1] Additional factors influencing wettability are considered in section VII.

Foliar absorption of pesticides is markedly influenced by both structure and quantity of surface wax just as is wettability. Also, an inverse relationship apparently exists between overall cuticle thickness and pesticide absorption. This relationship, strangely enough, is considerably stronger than any correlation which may exist between cuticle thickness and either surface wetting or transpirational water loss. These interrelationships have been previously discussed (HULL 1964 a); additionally, MALISAUSKIENE (1964) found that among 12 herbaceous and 12 woody species, a direct relationship exists between susceptibility to 2,4-D and both thinness of the cuticle and number of open stomata. HOEHNE and WASICKY (1950) had previously shown a similar type of inverse correlation between cuticle thickness and susceptibility to the sodium salt of 2,4-D. Coffee, one of the species with a thin cuticle, was most readily injured.

In contrast to the relative lack of correlation between water loss and overall cuticle thickness (often studied as a function of leaf age), the surface wax does play a significant role in minimizing such loss. MACHADO (1958), who has made comprehensive studies of the epidermal ultrastructure and waxy covering of several species of South American palms, expresses doubt as to the xeromorphic significace of external wax deposits whenever such deposits are scattered; however, he believes that the continuous, flexible, adhesive wax film that exists on leaves of the carnauba palm and the upper surface of leaflets from the licuri palm (*Syagrus coronata*) could be quite effective in reducing water passage. Conclusive proof of the "waterproofing" effect of this wax film has been obtained most convincingly in fruit cuticles. HORROCKS (1964) observed that when the cuticle was isolated from apple skin and the underlying cellular material removed enzymatically, permeability of the membrane was increased 30- to 70-fold (depending upon variety) if the surface wax had also been removed with hot chloroform. In following the moisture loss from intact grape berries, RADLER (1965) found such loss to be markedly increased if wax was removed with a suitable solvent; peeling the fruit completely yielded only a slightly greater drying rate. GRNCAREVIC and RADLER (1967) studied evaporation of water through a plastic membrane coated with 30 to 70 μg./cm.^{2} of wax from grape berries or fractions thereof. They discovered that the hydrocarbon, alcohol, and aldehyde fractions were the active components which prevented water loss—their effect was identical to that of the complete wax or to mineral paraffin wax. The principal constituent of the cuticle wax, the triterpene oleanolic acid, had no effect on evaporation, nor did free docosanoic acid. A mixture of the C_{24} and C_{26} free fatty acid constituents of the wax reduced evaporation only slightly in this artificial system. BUKOVAC and NORRIS (1967) also found that water penetration through fractionated components of the leaf wax from *Plantago major* showed marked differences among the various components, suggesting that chemical composition as well as overall quantity may be of importance in affecting such penetration.

An additional and interesting role of the fatty and waxy components of cuticle has been presented by MULLER (1965). By gas chromatography, he found that air in the vicinity of branches of *Salvia leucophylla* and *S. mellifera* contained two phytotoxic terpenes, tentatively identified as cineole and camphor. Those from the former of the two sages were further shown to be highly soluble in hard paraffin, thus suggesting that their mode of entry into other plants (and resultant toxicity) may be via the surface cuticle and the cuticular lining of the substomatal chambers and intercellular spaces.

5. Fungistatic and other properties of the epicuticular wax.—Since the fungistatic properties of foliar wax have been discussed by MARTIN (1964), only recent or additional findings will be considered here. MARTIN *et al.* (1957) first reported that when the ether extract of apple leaves was fractionated into its various components, the acidic fraction extracted from the waxes with dilute potassium hydroxide reduced germination of apple mildew (*Podosphaera leucotricha*) conidia when sprayed on the leaves before inoculation. A chloroform-soluble material was also found in the cuticle of *Ginkgo biloba* leaves by JOHNSTON and SPROSTON (1965) which reduced spore germination and germ-tube growth of certain fungi. They believed this substance to be responsible for at least part of the resistance of ginkgo leaves to penetration by common pathogenic organisms. RAM (1962), investigating the fungistatic action of tea leaf wax, observed that both spore germination and germ-tube growth of *Pestalotia theae* were a function of concentration of the petroleum ether extract. Germination on films made from 0.04 to 0.2 percent solutions was markedly stimulated. Outside of this range it was still stimulated to some extent, and at one percent it was inhibited. Partitioning of the extract against water and subsequent paper chromatography of the concentrate demonstrated two ninhydrin-positive substances which strongly stimulated germination, and one ninhydrin-negative compound which was somewhat stimulatory. Of further interest is the recent finding of EPTON and DEVERALL (1968) that the relative resistance of bean leaves to the halo-blight disease caused by *Pseudomonas phaseolicola* may be related to the activities of the respective lipase systems which liberate linolenic acid. Such liberation apparently occurs from two fractions of lipids, each representing less than two percent of the dry weight of the leaves, and tentatively identified as galactosyl diglycerides. In contrast to the fungistatic agents found in some cuticular waxes, MARTIN *et al.* (1966) were unable to detect any active constituents from the surface wax of mature citrus lime leaves, which are resistant to *Gloeosporium limetticola*, a fungus responsible for the withertip disease. However, a dimethoxy furocoumarin, isopimpinellin, isolated from cellular components, was highly toxic. Ester and acidic fractions obtained from both young susceptible and mature resistant leaves also showed some activity toward this fungus.

The fungistatic action of foliar sprays of Bordeaux and Burgundy mixtures is apparently achieved through solubilization of the copper on the leaf surface. MARTIN and SOMERS (1957) demonstrated that such solubilization is brought about by water-soluble acids from leaf wax. The action occurred when the copper compounds were sprayed directly on leaves, as well as when they were sprayed on petri dishes and then followed with the acids. CRAFTS (1961) has obtained evidence that short chain esters of 2,4-D may hydrolyze during foliar penetration, thus liberating the 2,4-D ion for movement via an aqueous route to the symplast. SZABO (1963) additionally demonstrated that both the butoxyethanol and the propyleneglycol ether esters of 2,4-D are hydrolyzed on the leaf surface of bean and maize plants, as well as within the plant itself. It thus appears that the cuticular wax, or some constituent(s) thereof, provides a microchemical environment capable of causing solubilization, hydrolysis, and possibly other reactions, even before the penetrant has reached the cutin.

Two additional interesting charactreistics of surface wax include its growth regulatory activity and its ability to minimize heat build-up in the leaf blade. VLITOS and CUTLER (1960) observed that wax extracted from the rings surrounding the base of each node on sugarcane stalks was capable of inducing a four-fold increase in the elongation of *Avena* sections, even at the extreme dilution of 0.0005 p.p.m. By means of a thermistor-thermometer STEINHUBEL (1967) investigated the temperature regimes of leaves from several plant species including the needles of two conifers. Following solar irradiation or irradiation from an artificial heat source, temperatures of both intact leaves and leaves with the wax removed demonstrated the distinct protective effect from over-heating exerted by the waxy covering. This occurred in spite of the cooling effect of increased transpiration which was observed in leaves with wax removed.

b) *Cutin*

Although this section is limited principally to cutin, some portions involve its interrelationship with the other cuticular constituents, e.g., discussions on structure, environmental factors affecting development, light transmissive properties, etc. Such information was placed in this section whenever it was felt that cutin was the most important, if not the sole, constituent under consideration.

1. **Chemical and structural characteristics.**—Recent advances in instrumentation have added significantly to our knowledge of this complex biopolymer which makes up a variable but oftentimes major por-

tion of the entire cuticle—as a layer immediately underlying the surface wax coating. The chemical properties of cutin have been discussed by BAKER *et al.* (1964), CRAFTS and FOY (1962), MADER (1958), and MARTIN (1964 and 1966). A particularly comprehensive treatise on cutin biosynthesis is that of LINSKENS *et al.* (1965), which includes the extensive work of Heinen and his colleagues on the role of such enzymes as cutin-esterase and carboxycutin-peroxidase. The enzymatic degradation of cutin has been discussed by GOODMAN *et al.* (1967). Remarks here will consequently be limited to additional findings which have been reported. Terminology will be used according to that suggested by ESAU (1965 a), namely, "cutinization" as being an impregnation of the cell walls with cutin, and "cuticularization" as being the surface adcrustation of cutin in forming the cuticle.

Cutin is now considered as a polyester of long-chain fatty acids and other substances, including hydroxylated and both mono- and dicarboxylic acids. The hydroxyl groups may be terminal or intermediate on the chain, but the nature of the intermediate binding remains obscure. Work suggesting single chains to be linked by peroxide groups thereby forming double- and triple-chain units, is reviewed by LINSKENS *et al.* (1965). Since a portion of the polar groups remain free during polymerization, cutin is semi-lipophilic in nature. It thus has the ability to swell in the presence of moisture and to permit transpiration, as well as allowing absorption of water soluble substances. Evidence recently reviewed suggests that cutin precursors migrate through the cellulose layer of the outer epidermal wall in the form of minute droplets, and thence polymerize into cutin under the influence of oxygen. In ultrastructural studies of the apical epidermal cells of the *Avena* coleoptile, O'BRIEN (1967) has in fact observed large cytoplasmic globules, up to two μ in diameter, which could be distinguished from the vacuole by the absence of a limiting membrane. The osmiophilic nature of these droplets mirrored that of the cuticle matrix, *i.e.*, the electron density of both droplets and cuticle was low in some varieties but extremely high for both constituents in other varieties, thus suggesting their possible function as a cutin precursor.

The studies of CRISP (1965) have demonstrated that polymerization of procutin to cutin, especially the formation of peroxide linkages, is enhanced by ultraviolet irradiation. Following degradation of the cutin from *Agave americana* to its monomers, he determined the latter to consist of a mixture of hydroxy-fatty acids ranging in chain length from tridecanoic to octadecanoic. The major acid, constituting 50 percent of the original polymer, was identified as 9,10,18-trihydroxyoctadecanoic acid; 17 additional acids were characterized by gas chromatography. Three types of bonding were determined in cutin, and included ester, alkylperoxide, and ether linkages, the ratio of their frequency being 7:2:0.2, respectively. In studies of long-chain hydroxy acids of apple fruit cutin, EGLINTON and HUNNEMAN (1968) recently identified the same (above) octadecanoic acid, in both *threo* and *erythro* configurations. They also characterized four additional principal C_{16} and C_{18} acids in the cutin. Also working with apples, HUELIN (1959) prepared cutin by enzymatic dissolution of the epidermis with snail gut extract. On alkaline hydrolysis the cutin yielded a water-soluble fraction containing carbohydrates (arabinose, galactose, and ribose), as well as galacturonic acid and an insoluble fraction containing protein. Infrared spectral examination of the cutin showed no change in number of OH or C=O groups during storage, indicating no further polymerization; if this occurred it should be due to the formation of ester linkages, which would be accompanied by a decrease in number of free OH and C=O groups.

The structural relationship of cutin to the other cuticular components, namely the wax, pectin, and cellulose, has been considered in detail by CRAFTS and FOY (1962), VAN OVERBEEK (1956), and others. From the considerable research accomplished to date on this subject, perhaps the most striking finding is the extreme variation among species, with respect not only to the thickness of the individual layers of the four constituents, but also to the extent with which they are integrated or intercalated one with the other. As noted by FREY-WYSSLING and MÜHLETHALER (1965), polarized light studies show the cutin to be almost isotropic, and under the electron microscope it appears virtually structureless. Starting from the surface, and with reference to the tangential direction of the epidermal cells, the outer portion of the cuticular layer shows varying degrees of birefringence, depending upon the content of surface wax and wax intercalated with the cutin. Next come the layers of cutin and pectin, both almost isotropic, and finally the epidermal cell wall, the cellulose of which is positively birefringent. The relative extent of each of these four layers has been beautifully demonstrated by SITTE and RENNIER (1963) in their polarized light studies of a large number of plant species. By extracting the waxy constituents, they observed changes in birefringence; they also plotted change in birefringence as a function of distance from the cuticular surface. Thus, the abrupt or gradual change in gradient density of all cuticular constituents is accurately depicted. As mentioned, it is unbelievably complex and variable among the different species. Utilizing this technique, in combination with fluorescence microscopy and cytochemistry, HÜLSBRUCH (1966) extensively studied cuticularization in a single species, *Ilex integra*. From a primary cuticularized wall 0.5 μ thick enclosing a 0.2 μ cellulose layer, she followed development of both leaves and stems until the entire epidermal wall reached a thickness of about 20 μ. At an intermediate developmental stage the thickened and polymerized cuticle was underlain by an isotropic homogeneous wall. When saponified with ethanolic potassium hydroxide, a system of diminutive tubes was observed to grow upward in a radial direction from the cellulose wall, occupying the isotropic wall. Wax apparently deposited on the sides of these tubules initiated a positive birefringence. In other portions of the wall, and depending upon development, zones of negative anisotropy were found. We have abserved similar tubules in the mature cuticular walls of *Prosopis juliflora* leaflets, without saponification, but when stained with a six-dye schedule (SHELLHORN and HULL 1961). In ultra-structural studies of the epidermis near the apex of the *Avena* coleop-tile, O'BRIEN (1967) showed the bulk of the cuticle to consist of an apparently structureless matrix through which ramified a reticulum of electron-dense fibrillar material, ranging in thickness from about 500 A down to the limit of resolution. The fibrils stopped just short of the cuticle surface, but were concentrated at the base of the cuticle; the latter region corresponded to a very dark layer as viewed by light microscopy, when the section was fixed in glutaraldehyde/osmium tetroxide and stained with methylene blue/azure II.

2. **Environmental factors influencing cuticular development, and light transmissive characteristics.**—Relatively few experiments have been designed to evaluate the effect of environment on the develop-ment of cuticle. MARTIN (1964) discusses some. It also seems necessary to include some of the earlier literature in order to obtain any kind of an overall picture of this subject. Although development of must cuticu-lar constituents is no doubt environmentally controlled to some extent, probably the cutin itself is affected somewhat more than the other com-ponents, particularly with respect to the sheer thickness of the layer which develops.

Some research suggests that certain factors are relatively unimpor-tant in affecting cuticular development. In studying the foliar anatomy of normal and water-stressed *Pelargonium zonale* seedlings, AMER and WILLIAMS (1958) noted that although three or four rows of palisade cells occurred in the stressed plants as compared to only two in normal plants, there was no significant difference in cuticle thickness between the two groups. In fact, both the upper and lower epidermal cell walls from stressed plants were somewhat thinner than the controls, although they appeared to be of a denser structure.

Temperatures to which roots are subjected apparently have no more effect on cuticular development than does available moisture. CORDS (1966) grew *Chrysothamnus nauseosus* seedlings in the greenhouse with their roots subjected to temperatures of 45°, 60°, 75°, and 90°F. Thickness of the leaf cuticle averaged about four μ at all temperatures, which was equivalent to the average thickness on previously examined field plants. Differential susceptibility of the various temperature groups to 2,4-D could apparently not be ascribed to quantitative cuticu-lar development. DONOHO et al. (1961) determined the extent of cuticu-lar development on leaves of apple and peach seedlings grown one year under controlled environments. Although variations in foliar absorp-tion of ^{14}C-NAA were noted among seedlings grown at different tem-peratures (60° and 70°F.) and different relative humidities (45 and over 90 percent), these could not be ascribed to cuticular development (determined by weight of cuticle), which was approximately equivalent under all environmental regimes. To test the possibility that storage rot in apple fruit might be associated with mineral deficiencies, BATT and MARTIN (1966) quantitatively examined the fruit cuticle of trees subjected to a range of fertilizer treatments. No significant differences could be established in total cuticle weight/unit surface, nor in total cuticular wax, surface wax, embedded wax, methanol soluble substances, acid and alkali soluble substances, cellulose, and cutin. Paraffins, esters, and acids of the surface wax were likewise unaffected by nutritional treatment.

In contrast to these reports suggesting a lack of correlation between certain environmental factors and cuticular development, there are some studies indicating that some relationships do exist. WILKINSON (1966 a) found that stems grown from cuttings of *Tamarix pentandra* developed a mean cuticle thickness of 6.4 μ when on a 14-hour photoperiod. Grown on either a shorter or longer photoperiod, cuticle development was only half or less of that with the 14-hour photoperiod. By growing numerous plant species in the humid atmosphere of bell jars, CUNZE (1926) noted that the majority developed a thinner waxy cuticle than did comparable plants under normal growing conditions, and others, namely Crassulaceae, had a normal waxy cuticle.

We have observed a very marked difference between the cuticular development of greenhouse-grown *Prosopis juliflora* seedlings and that of field plants, cuticles of the former generally being less than one μ thick, and those of the latter sometimes as much as 20 μ. Whether this difference might partially explain variations in herbicidal response sometimes noted between the two plant groups has been a speculative question for some time (HULL 1958). Cuticular development of the foliage of this deciduous plant in the field does not appear to be a function of plant size; also, it is unaffected by additional irrigation applied during the dry spring and early summer months which normally occur in the Sonoran Desert, its major habitat. Further (HULL 1964 a), seedlings were germinated and grown outdoors in the spring, and simultaneously in the greenhouse, which was maintained as closely as possible to the outdoor temperatures. The greenhouse plants, grown in vermiculite and irrigated with 65 percent Hoagland's solution, did not develop a cuticle of one μ thickness, even after 8.5 months. Seedlings grown in a similar manner, but outdoors, developed no more cuticle than did the greenhouse plants during the first few months. However, when they were grown in soil outdoors *and* allowed to reach an age of one year, the second season's leaves did form a moderately thick (four to five μ) cuticle. These studies emphasize the complex interaction of environmental variables which may influence cuticular development, including growing under glass, overwintering, and the possible effect of microelements which may be in the soil but were apparently absent in the Hoagland's-vermiculite substrate. To establish the effect of solar filtration by glass *per se* on cuticle development of a mature tree, a final experiment was carried out on a large field specimen of *Prosopis* (HULL and SHELLHORN 1966). Several weeks before budbreak in the spring, the terminal portion of a south-facing branch was placed in a large ventilated aquarium with the open baffled top facing north in such a way that all sunlight impinging on the enclosed foliage penetrated the glass. Microscopic examination of numerous leaflets from within the chamber 13 weeks after budbreak showed a mean cuticle thickness of five μ. Leaflets taken simultaneously from a branch immediately adjacent had eight μ cuticles. The significantly reduced cuticular thickness on leaflets within the chamber was probably due to near elimination of the shorter wavelength ultraviolet radiation. However, in spite of ventilation, both temperature and relative humidity (which were monitored inside and outside the chamber) averaged several units higher inside, and must also be taken into consideration.

SCHROETER (1923), studying the plant life of the Alps, found that the leaves of *Salix retusa* growing at 2,500 meters had thick leaves with two rows of palisade cells, small intercellular spaces in the mesophyll, and an extra layer of hypodermal cells under the lower epidermal cells. The cuticle was thick. When a plant collected from the mountains in the autumn was subsequently cultivated in the humid air of a bell jar, the leaves which developed had only a single palisade layer, a very loose cellular structure with large intercellular spaces, and only a very thin cuticle. When grown in the absence of ultraviolet light, VAN DER VEEN (1960) noted that leaves subsequently irradiated at 253.7 nm. from a bactericidal lamp were rapidly damaged and soon abscised. Newly formed leaves developed during ultraviolet treatment were hardier, smaller, and usually hairier. They also had a thicker cuticle, and were thus better protected from the harmful effects of the ultraviolet radiation.

According to LOCKHART and FRANZGROTE (1961), the cuticle of leaves and stems transmits relatively little ultraviolet of 275 nm. and is essentially opaque to the longer wavelengths of 350 to 400 nm. Celluloses and hemicelluloses are relatively transparent to the latter wavelengths, whereas the epidermal cell sap of some species sometimes absorbs strongly in this region, apparently due to tannins. Ultraviolet is also highly reflected by tissues with thick waxy coatings, and its transmission by waxes decreases strongly between 400 and 300 nm. LAUTENSCHLAGER-FLEURY (1955) observed somewhat similar light transmissive properties in *Vicia faba* leaves. Isolated, dried, epidermal strips were examined for transmittancy for the entire region from far-ultraviolet through infrared. Between 450 and 800 nm. a 70 percent transmittancy was obtained, but a strongly decreased transmittancy was noted in the range of 350 and 280 nm. Transmission of ultraviolet by the upper epidermis was less than that by the lower, but no differences were discernible with visible and infrared spectra. The work suggested that cutin does not have an ultraviolet absorption maximum, but absorbs 20 to 60 percent of the ultraviolet from 250 to 400 nm. No variations due to season or time-of-day were found. The lack of an absorption peak is probably due to the fact that cutin has few or no double bonds. However, as pointed out by FREY-WYSSLING and MÜHLETHALER (1965), although cutin is theoretically transparent to ultraviolet, it may contain (particularly in xerophytes) a pale yellow flavone pigment which would account for some absorption in this portion of the spectrum.

3. Fungistatic characteristics.—The cuticle's role in the defense against plant disease has been comprehensively reviewed by MARTIN (1964), and constituents of the surface wax which may be toxic to certain microorganisms have already been considered in the present discussion. Apparently the majority of such toxicants are confined to the surface wax, although some may be located more generally in the entire cuticle or even the cellular components of the leaf. Experimental technique in many cases precludes pinpointing the exact origin of some of these compounds. FISHER (1965) has isolated seven quercetin-3-glycosides, p-coumaric acid, p-coumaryl-quinic acid, chlorogenic acid, and phloridzin from apple fruit cuticle. He considered that the phenolic compounds could be fungicidal, although their concentration of less than one percent of the cuticle seemed too low for effectiveness in this respect. In spite of fungistatic constituents of this nature, pathogens do of course penetrate the cuticle and gain entrance to the underlying cells. Excellent ultrastructural details of the haustoria of *Erysiphe graminis* in the process of penetrating barley epidermal cells has recently been presented by BRACKER (1968). Some fungi apparently penetrate in rather specific patterns. For example, PREECE *et al.* (1967) found that in 90 to 92 percent of the cases, germinating conidia of *Erysiphe polygoni* on red clover leaves and *Peronospora parasitica* on cauliflower leaves formed appressoria in the junction areas between the anticlinal walls of adjoining epidermal cells. When leaf inoculations of the radish strain of *Pellicularia filamentosa* were made, KERR and FLENTJE (1957) noted that appressoria formation and fungal penetration occurred only where the cuticle was intact, and never through wounded tissue from which small strips of epidermis and cuticle had been removed. The cuticle in this case apparently acts as a fungal attractant, affecting both the attachment to and penetration of the host. The possibility cannot be ignored, however, that the buildup of a substance in the wounded tissue may occur which is inhibitory toward the pathogen. A final point of interest noted by MARTIN (1961) is the fact that infection of certain leaves with mildew or scab causes an increased deposit of surface waxes but decreased formation of cutin. In general, MARTIN (1964) considers that the cuticle does not offer a

great protection against pathogens, particularly when cutin weight is less than 0.1 mg./cm.^{2} of surface. Penetration via stomata or other natural openings is of course often possible.

4. Distribution and structural variations.—The internal cuticle lining the substomatal cavity has been described by Currier and Dybing (1959) and others. Chemically, it is probably more closely related to suberin than to cutin. In *Helleborus niger*, the ultrastructural studies of Huber *et al.* (1956) clearly differentiated such cuticular lining from the underlying cellulose of the guard and epidermal cell walls. It averaged about 0.15 μ in thickness. This is probably thinner than could have been estimated using the optical limits of the light microscope, where diffraction makes an accurate measurement difficult. A lipid layer is also known to exist on the leaf mesophyll cell surfaces adjoining intercellular spaces. Scott (1964) has observed this layer to be relatively thick in xerophytes and thin in hydrophytes. She considers its distribution in such spaces essentially to be a wound reaction—the wounding of the protoplast occurring during cell expansion, when the interconnecting plasmodesmata are broken. The protoplast presumably reacts by elaboration of wound hormones and lipid precursors. Air is essential for such deposition, which becomes thicker with growth. Another place where a cutin-like substance may exist is on the surface of plant roots. Martin and Fisher (1965) found that hydrolysis of the outermost tissues of roots of several plant species yielded mixtures of hydroxy-fatty acids. Although diversity was noted among species, and also between the acids isolated from leaf cuticle and root hyrolysate, there were several acids common to both the leaf and root of *Chlorophytum elatum*, 18-hydroxyoctadecanoic acid and three unknown acids. The findings suggest the root surface material to be more akin to suberin than to cutin.

Returning to the cutin of the leaf epidermis, various anomalies are sometimes observed. Anderson (1934) described an inner cutinized zone in the outer epidermal cell wall of *Clivia nobilis*, which occurred in addition to the outer one. Norris and Bukovac (1968) reported that the lower epidermis of the pear leaf was frequently cutinized on the inner surface of the inner epidermal cell walls, where it varied from isolated patches to an almost continuous layer such as occurs on the outer surface. Whenever exposed to air spaces the internal cutinization often developed an extremely corrugated surface. In fluorescence microscope investigations of the *Prosopis juliflora* leaflet we have observed a similar cutin-like material along the inner periclinal epidermal walls of both upper and lower epidermises (Hull 1964 a). During the seasonal development of *Tamarix pentandra*, a progressive subepidermal deposition of Sudan IV-staining substances was noted by Wilkinson (1966 b) to occur in the cladophylls. This "subepidermal cuticularization" reached a maximum of about 50 μ by late June and enveloped four to five layers of cells. It could be differentiated from the cuticle itself, which finally reached a thickness of about eight μ. Leaves did not develop the Sudan IV-staining substances. Secondary cutinaceous or lipoidal layers of this nature could well be an additional barrier to the penetration of pesticides, particularly when their polarity or ionization characteristics are incompatible with those of the penetrating substance.

Variation in weight/unit area of cuticular membranes and of surface wax has been investigated by BAKER and MARTIN (1967) for the families Saxifragaceae, Rosaceae, and Leguminosae. They observed wide differences in both components, even within the species of one genus. Variations in the relative quantity of four hydroxy-fatty acids in the cutin also existed, although the magnitude of this variation was somewhat less. ROUX LOPEZ (1964) has described some epidermal modifications of various Mexican xerophytes; for example, she noted the presence of a calcareous layer in the cuticle of certain species of *Opuntia, Agave, Yucca, Hesperaloe* and *Acacia.* In electron microscope studies of *Lilium candidum,* MAIER (1968) found delicately branched dendritic structures which penetrated the cuticular layer along its inner border. The structures were suggestive of excretions of a lipoidal-like cutinaceous material, the periphery of which had become strongly impregnated with silver during fixation. Since they were confined to the cuticular layer outside of the outer epidermal cell wall, they apparently bore no relationship to structures such as ectodesmata which are confined to the wall.

A final item of interest involves the gummy layer which covers the cuticle of tobacco leaves. According to MICHIE and REID (1968), this aromatic material or its precursors apparently originate, at least partially, in the trichomes. The polar lipids were found to consist of a complex mixture of non-volatile terpenes, which could be chromatographically separated. The components consisted chiefly of terpene diols, in which the α- and β-isomers of 12-isopropyl-1,5,9-trimethyl-4,8,13-cyclotetradecatriene-1,3-diol predominated. It was found that when either peeled leaf cuticle containing trichomes, or trichomes only (removed from the stem) were suspended in a buffer containing ^{14}C-acetate or ^{14}C-mevalonate, the synthesis of labeled terpenoids occurred.

5. Cuticular pores.—The existence of discrete microscopic or submicroscopic pores in the cutin which might serve as portals for water exchange or transport of wax precursors has long been a subject of conjecture. CRAFTS (1961) has noted that substructurally, the cuticle is perforated with micropores which are more or less filled with an aqueous phase, depending upon the environmental conditions to which the plant has been subjected. An aqueous continuum of this nature would of course seem necessary for penetration of water soluble solutes. However, the great majority of microscope observations, including ultrastructural studies of both cross sections and surface replications, has failed to show the presence of any larger discrete structures which might be termed cuticular pores. These negative reports have been recently reviewed and will not be repeated here. Dous (1927) observed with incident light microscopy what were apparently threads of wax excretions from pores on the surface of *Primula kewensis* leaves. In *Agave* such pores appeared to completely perforate the cuticle, and he assumed they served for a similar type of excretion. More recently, CRISP (1965) failed to observe cuticular pores in his ultrastructural studies of *Agave americana* and *Plantago major*. He did, however, note the presence of slime pores in *Elodea densa* cutin which extended from the cuticular-pectin layer into the polymeric cutin. They were considered to be part of a slime transport system in this water plant. The ultrastructural studies of SCOTT (1966) on several plant species suggested the presence of cuticular pores. Although pores were not visible in the shrunken isolated cuticles examined, their previous loci were indicated by the numerous thinner areas visible in the membranes.

By using a low-viscosity plastic for his first stage replication in combination with vacuum infiltration, HALL (1967 a) was able to confirm the presence of cuticular pores on the leaves of all six species examined; also on apple fruits. This negative replication had spikes where the pores had existed, the length of which could be determined by metallic shadowing at a known angle. However, it is not known that this represents the total depth of the pores, because of possible limitations in the replication technique. In species with sparse surface wax deposits it was possible to identify pores and canals under a single deposit. No pores could be detected on the inner surface of apple or clover cuticle after its isolation with pectinase, apparently because of its great roughness.

The presence of pores in the cuticles of various specialized glands has also been demonstrated by electron microscopy. SCHNEPF (1963) prepared cross sections of the epidermal outer walls of gland cells of *Drosophyllum lusitanicum* 14.5 hours after surface feeding with albumin. He found that although the cuticle was partially disintegrated or dissolved in some portions, pores were visible in other parts which led partially or entirely through the cuticle to the epidermal cell wall. The albumen apparently reached the wall. It appeared diffusely distributed therein, as well as in the vacuoles of the epidermal cells. THOMSON and LIU (1967) observed that the outer surface of the salt-secreting gland of *Tamarix aphylla* is covered with an apparently cuticular layer up to 1.5 μ thick. In the central area where this layer is invaginated amongst the six secretory cells, it contains occasional electron transparent areas about 0.5 μ long, and 0.1 to 0.2 μ in diameter. These appear to be pores, and apparently come within about 0.1 μ of the surface. With increased turgor of the secretory cells, the salt brine is perhaps forced along the cell wall and out these pores in the top of the gland.

6. Resistance to decomposition.—Because of the exceptional resistance which leaf cuticle has to chemical decomposition, it is often observed in fossilized form. This characteristic has been documented by a number of reports, only a few of which are mentioned here. There is apparently considerable variation among the cuticles of different species, as to their potential to become fossilized. STEWART and FOLLETT (1966) used electron microscopy to study the progressive decomposition of cuticular surfaces and noted that relative decomposition rate during peat formation was a function both of plant species and of depth in the peat. Some species showed various degrees of surface etching or loss in definition of surface features; others did not, thus suggesting a range of fossilization potential. Studying palm leaf fossil material from Eocene deposits in southeastern North America, DILCHER (1968) found good cuticle preservation. Leaf fragments from this epoch could be readily related by cuticular analysis to more complete leaves which were obtained. Finally, the cuticle's resistance to breakdown is evidenced by its recovery from the rumen, stomach, and fecal material of grazing animals and subsequent identification with the cuticles of known plants. This has proven useful in identifying the plants grazed, as has been demonstrated by MARTIN (1955) with sheep, and by other investigators.

c) *Cuticular penetration and adsorption, including studies with isolated membranes*

Experiments involving the use of chemically or enzymatically isolated cuticular membranes have added considerably to our knowledge of adsorption and penetration phenomena. Details of techniques used in cuticle isolation have been thoroughly discussed by FRANKE (1967 a). Recently HOLLOWAY and BAKER (1968) found that a solution containing one g. of zinc chloride in 1.7 ml. of concentrated hydrochloric acid could cleanly separate cuticular membranes within two to 12 hours. The solution was effective on a wider range of species than were either the ammonium oxalate/oxalic acid or the pectic enzyme methods. Experiments utilizing isolated membranes have been criticized on the grounds that such isolation is likely to induce certain chemical alterations of the membrane, as well as subjecting it to stresses which could cause physical imperfections. However, most workers inspect the membranes under high magnification and verify their condition with nonpenetrating marker dyes; any of those which are not apparently perfect are discarded. Perhaps the most convincing evidence of an almost complete lack of alteration during isolation comes from the recent work of NORRIS and BUKOVAC (1968). They examined both intact and enzymatically isolated cuticles of pear leaves with normal and polarized light, and also by electron microscopy. The staining properties with Sudan III and IV and with ruthenium red were similar in both types, as were the intensity and location of bire-fringence within the cuticle and the structure of the platelet-like wax formations on its surface. The original spatial configuration of the cuticle was maintained following isolation, even within the substo-matal chambers. Continuity of such cuticle with that of the outer sur-face was thus conclusively demonstrated.

The previously discussed rôle of waxes in foliar penetration has also been studied with enzymatically isolated cuticles (BUKOVAC and NORRIS 1967). It was noted that successive removal of the surface wax from pear leaf cuticles increased both penetration and adsorption of NAA; however, only slight additional increases resulted following ex-traction of the embedded waxes. When the wax was plated onto de-waxed cuticles both adsorption and penetration were again reduced, proportionately to the quantity applied. Concentration effects and kinetic data suggested that penetration through the isolated mem-branes was primarily a diffusion process. BUKOVAC and NORRIS (1968) also found that binding of NAA and NAAm increased linearly with concentration. Binding of NAA was greater at pH three and four than at five or six, whereas no significant change was observed with NAAm over an extended pH range. Also, NAA was bound to a greater extent by the lower cuticular membrane than by the upper, and more by isolated membranes than by intact. NAAm showed no significant dif-ferences in this respect. However, both compounds were bound more extensively by young expanding leaves than by more mature leaves. Binding apparently involved sorption into some component of the cuticular membrane, with electrostatic forces playing a minor role. Microautoradiography following application of ^{14}C-3-chlorophenoxy-α-propionic acid to isolated cuticular membranes failed to demonstrate localization of activity which could be associated with any visual mor-phological features of the membranes, but disclosed a considerably greater binding by lower than by upper membranes, as observed for NAA. In diffusion studies with the isolated lower cuticular membrane of apple leaves KAMIMURA and GOODMAN (1964 a) found ^{14}C-labeled leucine to be more mobile than glucose, urea, benzoic acid, or aspartic acid. However, less than two percent of the radioactivity available for diffusion actually moved through the membrane. Of the antibiotics studied, chloramphenicol and streptomycin were more diffusible than any of four tetracyclines, yet less than 0.2 percent of the available activity of even the most mobile antibiotics penetrated the membrane.

Additional information of interest has been obtained through the use of isolated cuticular membranes for studies of inorganic ion penetra-tion. HAILE-MARIAM and WITTWER (1965), using enzymatically isolated membranes of the upper astomatous and lower stomatous surfaces of *Euonymus japonicus*, observed the following relative penetration rates: Cs=Rb>K>Na. Rate was thus inversely correlated with size of the hydrated ion. As would be expected, penetration was greater and size of the hydrated ions was of less significance for the stomatous than for the astomatous cuticles. Also using *E. japonicus,* and laurel in addition, SILVA FERNANDES (1965 b) found no diffusion of copper from copper acetate or sulfate solutions, or of mercury from a phenylmercuric acetate solution, though stomata-free leaf membranes contained 0.1 mg./ cm.² or more of cutin. However, diffusion occurred through membranes containing stomata. Wax also proved to be an important barrier to the penetration of mercury through cuticular disks of apple fruits. The effect was dependent upon the wax composition, adsorption of mercury by the disks being increased by an increase in the percentage content of esters in the wax, by a rise in temperature, or by the presence of a wetting agent. The exceptionally rapid foliar absorption of urea has been a subject of interest for some time. YAMADA *et al.* (1965) have observed that at a concentration of 10 mM, urea additionally has the ability to enhance the penetration of rubidium and chloride ions through astomatous cuticular membranes enzymatically isolated from ripe tomato fruit. This urea-induced increase in permeability may partially explain reports of the increased effectiveness of nutrient sprays of phosphorous, manganese, and iron when used in combination with urea. However, KANNAN (1969) has just reported that urea reduced the penetration of ^{59}Fe, from either ferrous sulfate or a chelated form, through isolated cuticular membranes of tomato fruit and leaves of *E. japonicus.* Penetration of iron from the inorganic source was the more rapid. For further details regarding the mechanism of ion binding by isolated cuticular membranes, including the penetration-enhancing activity of urea, the review by FRANKE (1967 a) should be consulted.

III. The epidermis

a) *Surface configuration*

1. Taxonomic and developmental aspects.—The basic forms of the cuticular surface, which are in turn a function of the growth pattern of underlying epidermal cells, are becoming increasingly important as a taxonomic aid. A comprehensive monograph on this subject has been published by STACE (1965). The principal forms include lamellate, striate, and reticulate. Lamellate is most common and reticulate quite rare. The latter has been observed by RAO (1963), however, on the lower surface of *Hevea brasiliensis* leaves, where it forms a median ridge on each epidermal cell from which several arms with tapering ends originate. The median ridge is generally not present on subsidiary cells or guard cells. Even the young (1 × 2 cm.) leaves had the reticulate configuration on their lower surfaces. Macroscopic leaf morphology is often inadequate to characterize individual species as, for example, in the case of the genus *Vaccinium.* However, MUELLER (1966) demonstrated that the microscopic surface structure of leaves of this genus, when investigated by a plastic replica technique, showed a different pattern for each species. Year-to-year variation of pattern within an individual plant and also ontogenetic variation showed a generally consistent pattern within a given genotype. She did observe, however, that the pattern is under rather strong genetic control, and that with hybridization it breaks down, resulting in an appearance somewhat modified from the original.

The extent to which environment influences overall development of various surface features has been studied by SHARMA and DUNN (1968) with leaves of *Kalanchoe fedschenkoi*. Using cuttings from a single parent to maintain genetic constancy, they subjected the plants to ten different environmental conditions. Stomatal index, subsidiary cell pattern, stomatal development, smallest and largest stomatal sizes, and length-width ratio of the leaves were not modified by the environments studied. In contrast, stomatal frequency, absolute stomatal number, epidermal cell frequency, and leaf size and thickness were all readily modified.

2. Methods of studying.—Morphological features of the epidermis may be examined by means of plastic surface replications or epidermal strips. A modified technique for the preparation of epidermal strips which has proven particularly useful for leaves of Gramineae has been described by CLARKE (1960). The mounted strips clearly show the development of papillae, hairs, spines, and other features valuable in taxonomic, developmental, and genetic studies. Numerous types of plastics and waxes have been described for making epidermal imprints. One disadvantage of this method is that leaf hairs are generally not clearly recorded in the replica, and therefore an assessment of their contribution to water repellancy, or for taxonomic purposes, etc., may be lost. HORANIC and GARDNER (1967) found Rhoplex AC-33 (a viscous emulsion of acrylic polymers) particularly useful for epidermal imprints, since it is nontoxic to the plant and would thereby allow preparation of repeated imprints for studies of guard cell movement or of epidermal cell development. Studying the surface features of *Festuca pratensis* leaves, CHALLEN (1960) found the use of carnauba wax positive replicas particularly useful and demonstrated that macroscopic surface roughness was the chief factor in preventing wetting.

Extreme detail of the surface topography may be determined by means of interference microscopy, as described by LINSKENS (1966) and LINSKENS and KROES (1966). The method involves pressing a thin (0.07 mm.) plastic film, softened with acetone, to the leaf surface. After three to five minutes the replica is removed, placed in a Zehender chamber, and examined for interference with reflected light. Interval between the interference lines is dependent upon the nature of the phase adjacent to the replica and may vary from 0.54 to 5 μ. The great sensitivity of this technique makes it useful for quantitatively recording changes in the developmental pattern of the leaf surface under different physiological conditions, or as it may be affected, for example, by the application of pesticides. Using the method in combination with electron microscopy, LINSKENS and GELISSEN (1966) observed that the russeting in Golden Delicious apple fruits, which appears as star-shaped craters in the skin, actually consists of an alteration of the epidermal and cuticular layers, apparently caused by inhibition of cutin synthesis or polymerization, or by a shortage of lipid material. Under the indirect influence of pesticides, russeting could extend to a general pathological deviation.

b) *The cell wall*

1. Structure.—The chemical and physical composition of the plant cell wall, including the biosynthesis of its polysaccharides, has been thoroughly discussed by FREY-WYSSLING (1959), GOODMAN *et al.* (1967), MÜHLETHALER (1961 and 1967), ROELOFSEN (1959), ROGERS and PERKINS (1968), SETTERFIELD and BAYLEY (1961), SIEGEL (1962), and WILSON (1964). As described by MÜHLETHALER (1961), the cellulose layer often shows lamellation, interspersed with pectic substances. The amount of pectin per layer increases with distance from the lumen, and the boundary between the cellulose layer and the cutin may be sharply defined by a pectin sheet. Because of this intimate association of pectic substances with the cell wall, their role is generally included in the above and other reviews pertaining to the wall structure; they will be so considered here, even though pectin substances are sometimes pictured as a component of the cuticle. Pectin was perhaps first characterized as a cuticular constituent by GÉNEAU DE LAMARLIÈRE (1906), who demonstrated it with 1/5000 "rouge de Ruthenium." A solution of ruthenium sesquichloride with a little ammonia is still commonly used for identification of pectin. However, as pointed out by ROELOFSEN (1959), it is not specific and must be supplemented by observations on how the sections stain after removal of the pectin with ammonium oxalate. Since ruthenium red is a cation, it attacks not only the carboxyl groups of pectin, but also those of any cutin or wax prescursors which may be present. Pectin has also been demonstrated histochemically in ultrastructural studies. ALBERSHEIM and KILLIAS (1963) developed specific electron-dense stains for pectin and nucleic acids. Alkaline hydroxylamine, for example, formed a pectic hydroxamic acid complex with ferric ions, which gave excellent clarity and selectivity. In a study of pear leaf cuticle, NORRIS and BUKOVAC (1968) observed that pectic substances extended well into the cutin matrix, particularly over the anticlinal walls. However, they never reached the outer surface.

Considering cellulose, CZAJA (1962), in a series of superb polarized light photomicrographs, has added to our knowledge of the submicroscopic structure of the outer epidermal cell wall. He demonstrated how

monocots with relatively long epidermal cells have cellulose fibrils oriented parallel to the longitudinal axis of the leaf blade, whereas monocots with shorter epidermal cells have a fibril structure which is perpen-
dicular to the blade axis, as in maize, for example. Details of fibrillar
orientation in the outer periclinal wall were worked out for epidermal
cells above veins and in intercostal areas; and for cells of rectilinear,
sinuous, and round outline, as well as for guard cells, accessory cells,
and leaf hairs. The epidermal wall of the trigger hair podium on leaves
of Venus's-flytrap (*Dionaea muscipula*) has been shown by SIEVERS
(1968) to have an unusual ultrastructure. Numerous radially arranged
fibrils about two nm. in diameter were found inserted in the fibrillar
network of the outer epidermal wall beneath the cuticle. It was thought
that the structures increased the elasticity of the podium, allowing it
to undergo repeated bendings.

LEIGH and MATTHEWS (1963), in an ultrastructural examination of
leaves from four varieties of love grass (*Eragrostis curvula*), noted considerable variation in the density of osmium granulation in the outer
epidermal wall of the different varieties. In one, a wide band of osmiophilic granules occurred immediately beneath the cuticle in the epidermal wall. O'BRIEN (1967), studying the ultrastructure of epidermal
cells from the apex of *Avena* coleoptiles, also noted in osmium-fixed
material the presence of two principal layers in the outer periclinal
wall. The outer portion consisted of a finely-textured mosaic of electron-
dense granules and fibrils 30 to 80 A in diameter, along with another component of more electron-transparent granules and fibrils of
slightly larger dimension. The nature of the fibrillar material was not
certain, although polyuronides seemed a possibility. Osmiophilic cutin
cystoliths were also found embedded in the outer portion of the wall.
The inner wall region consisted essentially of a lamellated structure,
and had previously been shown (O'BRIEN 1965) to contain numerous
linearly-arranged vesicular elements. These elements, apparently described for the first time, varied in frequency among different *Avena*
species. The inner portion of the cell wall did not seem to be polysaccharide in nature, because of a reduced staining with the periodic
acid-Schiff's reagent (JENSEN 1962). Also studying the outer periclinal
wall of the *Avena* coleoptile epidermis, ROSSI and ARRIGONI (1965) reported a new type of lamellar structure which was embedded most
frequently in the inner non-cutinized portion of the wall. Orientation
was largely parallel to the cell surface and followed a regular pattern
of alternate layers of electron-transparent and electron-dense bands.
These lamellar bodies had a thickness of 200 to 300 A and were
bounded by a 30-to-40-A-thick membrane. Embedded in the outer
cutinized portion of the wall were some elliptical bodies surrounded
by a single membrane, 20 A thick, perhaps the "cutin cystoliths" described above by O'BRIEN. The interior of these bodies revealed thin
lamellae enclosed in an electron-transparent stroma. With unfavorable

growth conditions the frequency of the structures declined greatly. It was considered that they may be somehow involved in the growth and development of the cell wall.

There is now good evidence that a number of enzymes occur within the free space of higher plant tissues. In addition to any rôle these enzymes might play in biosynthesis of the cell wall, they could of course also act upon any exogenous substance being transported through the apoplast system, and as such are pertinent to this discussion. These enzymes include certain invertases, pectin methylesterase, uridinediphosphoglucose (UDPG) pyrophosphorylase, inorganic phosphatase, α-glycerophosphatase, UDPG fructose transglucosylase, cellobiase, ATPase, pyrophosphatase, nuclease, amylase, indoleacetic acid oxidase, and others. The enzymes perform various hydrolytic and synthetic functions, including, in some instances, a direct or indirect effect on the transport of organic and inorganic substances within the apoplast. Research relating to the occurrence and origin of enzymes specifically performing this latter function has been reviewed by SACHER (1964). In addition to enzymes, cell walls are known to contain proteins. Evidence for their presence, originally thought to be cytoplasmic contaminants, has been discussed by MÜHLETHALER (1967).

2. Plasmodesmata.—The important role played by plasmodesmata in the intercellular transport of numerous substances is well documented by research reports, and more recently by ultrastructural studies which generally substantiate this rôle. The subject has been discussed comprehensively by SCHUMACHER (1967 a), VOELLER et al. (1964), and several other reviews listed in the introduction. Although categorized here under the epidermis, some of the work described also pertains to plasmodesmata of parenchyma cells where the basic structure and function of these components are probably similar to those of the epidermis. The fine structure of the plasmodesmata of *Phalaris canariensis* has been examined by LÓPEZ-SÁEZ et al. (1966). They reported the outer membrane of the cylindrical structure to be continuous with the plasmalemma of the adjacent cells. Lying in the center of the cylinder thus formed (total diameter 400 to 500 A) was the "canaliculus". The latter, in transverse section, appeared as an inner black point surrounded by an outer dark ring, with a diameter of about 150 A. This inner tubule apparently consists of endoplasmic reticulum and does not constitute an intercellular canal, since it is filled with the unit membrane which forms its own wall. Between this wall and the outer cylinder, however, appeared an optically empty space which the authors termed the "intertubular gap," and which they thought could act as a connection between the hyaloplasms of the adjacent cells. A generally similar type of substructure was recently reported by ROBARDS (1968 a and b) for the plasmodesmata situated between differentiating ray cells of *Salix fragilis*. Double outer dark layers representing the plasmalemma were likewise continuous through the plasmodesmatal canal. A central rod of 40 A diameter also ran through each plasmodesma, and was surrounded by a tubule (suggested name, "desmotubule") of about 200 A diameter with a 50-A thick wall, to which the rod was attached with fine filaments. The tubule wall showed some substructure, probably consisting of 11 sub-units. At the ends of the plasmodesma the tubule was closely bound by the plasmalemma, suggesting in this case, no continuity of either cytoplasm or endoplasmic reticulum between adjacent cells. The endoplasmic reticulum, while closely approaching the plasmodesmata, was not continuous across them. The possibility was considered that the inner tubular structures may be nuclear spindle fibers.

The endoplasmic reticulum appears to pass completely through the plasmodesmata in some cases, as, for example, has been observed by DOLZMANN (1965) in the plasmodesmata between the dome cell and the first absorptive cells of the absorptive hairs of *Tillandsia usneoides*. According to ESAU (1965 a), some studies show no plasmodesmatal connection between guard cells and adjacent epidermal cells, although such connections have been demonstrated between guard cells and subsidiary cells. There has apparently been a divided opinion as to whether or not plasmodesmata exist between guard cells and subsidiary cells of the leaves of bean, tobacco, and *Datura stramonium*. That they do indeed exist has recently been confirmed by LITZ and KIMMINS (1968). In an ultrastructural study of 16 species of grasses BROWN and JOHNSON (1962) found no plasmodesmata at the ends of the guard cells where they are in contact, but observed that the wall between the two cells was incomplete. The protoplasts were partially confluent and common to both cells. In an electron microscope study of rye leaf stomata MIROSLAVOV (1966) likewise observed large perforations in the membrane between the broad end sections of the guard cells, allowing direct contact of their cytoplasm. This picture is apparently common in the grasses.

Of additional interest is the work of SPANSWICK and COSTERTON (1967) on the plasmodesmata between internodal cells of *Nitella translucens*. They determined the size and distribution of these structures by electron microscopy, and measured the specific electrical resistance of the node by means of intracellular microelectrodes. The micrographs suggested that the resistance should be smaller than the value actually obtained, by a factor of about 330. It was consequently assumed that there must be a restriction to the diffusion of ions in the plasmodesmata. In spite of the unexpectedly high resistance to ionic migration, it was considered that the plasmodesmata provide a far more efficient means of intercellular transport than the alternative of "secretion and reabsorption with its consequent wastage of energy and materials."

3. Ectodesmata.—The study of ectodesmata, including their structure, function, and even their detection, continues to be an area of zealous research. Ectodesmata are fine plasmodesmata-like structures,

but of an apparently non-plasmatic nature, and are found in the outer periclinal epidermal cell walls, primarily in leaves. They also occur in the walls of guard cells, the basal cells of trichomes, glandular hairs, and in epidermal cells surrounding these structures, including the anticlinal walls. Ectodesmata are generally considered partially or completely to penetrate the cell wall from the plasma membrane to the inner cuticular surface. However, the electron micrographs of *Helleborus niger* presented by HUBER *et al.* (1956) suggest that the ectodesmata do not always end at the cuticle, but often penetrate it nearly to the outer surface, perhaps even penetrating completely through in some cases. They are apparently absent in some plant species; their demonstration in others is indeed markedly influenced by environmental conditions, toxicants, and techniques of fixation for microscopic examination. Details of these factors, as well as structure and function in foliar absorption and cuticular excretion, have been discussed in a number of reviews cited in the introduction, with particularly comprehensive consideration by FRANKE (1964 and 1967 a), LINSKENS *et al.* (1965), NAPP-ZINN (1961), and VOELLER *et al.* (1964).

Ectodesmata may be visualized by fixation in Gilson solution which contains saturated mercuric chloride (SCHNEPF 1959). Although this technique had not previously been successful in demonstrating ectodesmata in Gramineae, FRANKE and PANIC (1967) have now been able to make these structures visible in leaves of wheat and maize by the addition of nitric acid to the basic fixation mixture. As in other plants, including gymnosperm and pteridophyte species, the previously described distribution of ectodesmata was observed and was characterized by the formation of rows along the anticlinal walls, crowding in guard cells, and a scattered distribution in periclinal walls. The demonstration of ectodesmata with the microscope is also dependent upon the wavelength and intensity of the light used. With interference filters and an intensity of 2,400 ergs/cm.^{2}/sec., SIEVERS (1960) found in *Primula veris, P. elatior,* and *Cortusa matthioli* that the visual rendition of these structures was suppressed with blue (430 nm.) and red (655 nm.) light, while green (540 nm.) and yellow (572 nm.) were only weakly effective.

Another factor, discussed in some detail by SUKHORUKOV and PLOTNIKOVA (1965), is the apparent disappearance of ectodesmata following the action of certain toxicants or pathogens. If the unfavorable stimulus is mild and subsequently removed, the ectodesmata again become demonstrable. However, with a more severe action, permanent degradation of the structures occurs and is accompanied by derangement and ultimate death of the protoplasts. The suggestion was advanced that the state of the ectodesmata may reflect the degree of infection from various diseases: for example, complete disappearance in zones of infection or intoxication. Thus, ectodesmata may enhance the plant's perception of environmental factors, their lability perhaps reducing the possibility of cellular injury from external pathogens or toxicants. Such a function would seem to imply the necessity of ectodesmata being plasmatic in nature, although Franke (1967 a) and others now consider this unlikely. An alternate implication might be the intimate association of ectodesmata with the protoplast, even though the former are non-plasmatic. A valuable adjunct to studies of this nature would be a method to measure slight pathological alterations in different areas and to correlate such alterations with the demonstrability of ectodesmata. The technique described by Burian (1964) would perhaps be applicable whereby pathological alteration of the epidermal cell sap can be monitored by delicate shifts in fluorescence from a normal bluish-gray to a brown secondary fluorescence in infected cells and by an intensification of the normal blue fluorescence of subepidermal tissue.

One of the difficulties regarding ectodesmata (Crafts and Foy 1962), is whether they correspond to "cuticular pores" of one type or another which have been reported by various workers. The majority of research reports, which are well documented in the reviews listed herein, suggest a general absence of any type of pore or channel within the outer epidermal cell wall which might serve, for example, as a pathway for outward transport of cutin or wax precursors. Leigh and Matthews (1963), studying the ultrastructure of *Eragrostis curvula* leaves, failed to observe any pores in the outer epidermal wall through which wax precursors could have been extruded during formation of the crystalline surface wax. Fisher *et al.* (1969), in their ultrastructural investigation of conifer needle outer epidermal cell walls, did not find any wax pores or ectodesmata in the nine species studied. Bancher *et al.* (1960) observed that isolated epidermal fragments of *Allium cepa* stained with neutral red and embedded in paraffin oil soon exhibited droplets of plasmolyzing sap on the cuticular surface. The droplets were more numerous over anticlinal than over periclinal walls and were assumed to be transported from the cell interior to the surface through ectodesmata. However, no pores or thin spots corresponding to the observed droplets were visible with the electron microscope. Maercker (1965 a) likewise failed to detect pits or pores in the outer epidermal walls of *Cocculus laurifolius* and *Camelia japonica,* and concluded that structures which may appear as such are simply artifacts related to the very undulating radial walls of the epidermal cells and which, in section, may evoke a deceptive appearance of a pore or pit.

Although ectodesmata have been observed in *Plantago major* leaves by light microscopy following fixation with Gilson solution (Franke 1964), the electron microscope studies of Crisp (1965) and Crisp *et al.* (1966) failed to establish any discrete structures in the leaves of this species which could be termed pores or ectodesmata. Instead, fan-shaped or dendritic-like patterns of mercuric chloride diffusion into the cell wall were detected and were termed Gilson Diffusion Patterns

(GDP). The GDP were not deposited uniformly within the cell wall, apparently due to preferential penetration through selective areas of the cuticular membrane. Also, since the reaction is dependent upon reduction of mercuric ions to the mercurous form, the aqueous solvent probably sweeps any necessary endogenous reductant along with its advancing front, where the concentration may become sufficiently high to show visible results of the reaction. Deposition of the relatively insoluble mercurous ions did not disrupt orientation of the cellulose microfibrils, except at advanced stages of deposition, where the cellulose was not displaced but appeared to be dissolved. The GDP were present more frequently on the upper leaf surface than on the lower and were not more numerous in and around the guard cells, as has been reported for ectodesmata (FRANKE 1964 and 1967 a). It was considered probable that the observed GDP were identical with the ectodesmata presented in the literature. Additional work on *Plantago, Primula, Viola,* and other species in which ectodesmata have been reported, has recently been carried out by BEASLEY and PHILPOT (1969) by means of light and dark field illumination and with polarizing and phase contrast microscopy. They found that diffusion patterns within the cell wall, such as those described above, could be demonstrated simply by application of mercuric chloride-saturated water to the leaf surface. They consequently proposed the term Crystalline Mercurial Patterns (CMP) for these structures, apparently identical to those described as ectodesmata. Since Gilson's solution was nonessential for demonstration of the patterns it was not included in the term; also, mercurial was used in preference to mercuric, mercurous, or such, because of uncertainty as to the chemistry of the final bound forms of mercury in the tissue. Considerable variation in the form of the CMP's was observed among species and within a single species. Sometimes only localized crystalline deposits occurred; more often a needle-like dendritic crystal growth or ribbon- or wedge-shaped forms were found. In *Gossypium hirsutum* the CMP completely penetrated the cell wall and were often observed to precipitate in a globular form at the interface of cell wall and cytoplasm. In agreement with the distribution of ectodesmata described by FRANKE (1967 a), there was often an abundance of CMP in the epidermal cell walls adjacent to or directly over main veins, although this was not always .true. Their frequency was also high in the foot cells of trichomes and precipitates were observed in stomatal apertures as well as within guard cells. None of the results obtained could logically explain the nonuniformity found in the appearance, location, and frequency of these CMP's. However, the workers felt that these patterns, as observed, offer no proof of preexisting pores or passageways in the outer epidermal cell wall.

In contrast to the numerous reports indicating the nonexistence of any type of channel within the cell wall, some research has suggested that such structures do exist. Perhaps most convincing of the earlier ultrastructural work is the investigation of SCHNEPF (1959) on *Primula veris* spp. *macrocalyx*. His electron micrographs (his figures 6 through 10) apparently depict ectodesmata in the wall of the leaf's upper epidermis, following fixation with Gilson solution. In contrast to plasmodesmata, they do not appear as thick canals but as fine fibrillary bundles in the microcapillaries of the cell wall. Although a considerably greater frequency of cuticular pores was observed than the 30 ectodesmata/100 μ^2 found, no pores or thin areas were visible in the cuticle directly over the ectodesmata. This makes it unlikely that the pores are extensions of the latter. SCOTT (1966) considers the possibility that any plasmatic constituent of the ectodesmata is probably retracted away from the cuticle as a result of the shock inevitable during preparation of the specimen for ultrastructural examination. In the *Bergenia* leaf, for example, she also observed a simultaneous shrinking of the elastic cuticle after it had previously been stretched by the cell turgor. No cuticular pores were visible, although the micrographs showed thin areas which were suggestive of loci of former plasmatic endings.

Using a modified freeze-etch technique in preparing epidermal specimens of *Trifolium repens* and other plants for electron microscopy, HALL (1967 b) demonstrated the presence of microchannels in the outer epidermal cell walls which apparently serve as a route for the transport of wax to the leaf surface. The channels had a central core usually six to ten nm. wide and an overall diameter of about 40 nm. Their orientation tended to follow that of the cellulose fibrils, and an abrupt change of direction was generally observed midway between the plasma membrane and cuticle—possibly at the junction of primary and secondary cell walls. Channels were visible entering the cell wall both from the plasma membrane and the cuticle, suggesting that they traversed the full width of the wall. The sharply demarcated boundary of these pits, their abrupt change of direction, and their small diameter (ectodesmata diameters are more in the range of 0.2 to 2 μ) all tend to negate the possibility that they are ectodesmata. This, however, brings up the question of terminology: should all channels in the outer epidermal cell wall be considered ectodesmata? Since the structures are apparently nonplasmatic, and the term holds no connotation to plasma, it would seem entirely appropriate. Another report of pores in the outer epidermal wall of *Allium cepa* scales comes from HÖLZL (1964). In ultrastructural studies of tangential sections fixed with potassium permanganate he observed a distinctly porous structure with the diameter of the roundish pores varying between ten and 100 nm. These submicroscopic structures, termed subectodesms, were considered to underlie the basic structure of the ectodesmata. The influence of environmental factors upon the demonstrability of ectodesmata was previously mentioned. Such influence may extend to the general microscopic or submicroscopic porosity of the wall, as suggested by MIROSLAVOV (1962), who studied the adaption of the leaf epidermis of six genera of grasses to arid conditions. Under such conditions the outer epidermal cell walls thickened, but they also became more porous, thus enabling a continued contact of the vital symplast with the external environment.

Little has been added to information relating to the function of ectodesmata in foliar absorption and transpiration since the review of FRANKE (1967 a). By autoradiography following application of ^{14}COOH-labeled α-aminoisobuteric acid to the leaves of *Vicia faba*, NEUMANN and JACOB (1968) demonstrated preferential absorption in the guard cells, particularly along the rear wall. Little penetration occurred through the stomatal aperture itself. A corresponding leaf preparation, not given butyric acid but fixed with the mercuric chloride reagent, showed a similar pattern of ectodesmatal distribution, suggesting preferential absorption by the ectodesmata as very likely. Their function as portals for transpirational water loss has also been suggested by the work of FRANKE (1967 b and c). Tritiated water taken up by the petioles of detached leaves of *Zantedeschia aethiopica* produced an image on the microautoradiogram indicating emergence from the guard cells, the so-called peristomatal transpiration. Silver grains appeared principally along the cuticular edges of stomatal pores, but not directly above the pores. This distribution also corresponded closely with that of the ectodesmata.

The possible relationship of ion binding sites within the cuticle to ectodesmatal distribution should be mentioned. YAMADA et al. (1966) found by microautoradiography that although no localization occurred in the binding of calcium and chloride ions on the outer or inner surfaces of isolated astomatous tomato fruit cuticle, such localization did occur on the cuticle of the stomatous green onion leaf. Binding sites for calcium on the outer surface and sodium on the inner surface were concentrated above the center of the periclinal cell walls in a line midway between the anticlinal walls. Concentration also occurred within the stomatal sac, the continuum of the cuticular membrane into the stomatal cavity. FRANKE (1968) has proposed two possibilities to explain binding of this nature: 1) penetration occurs through the entire cuticle, but only specific sites may be able to bind ions, and 2) penetration occurs *only* at the sites of binding. He subsequently investigated the distribution of ectodesmata in the corresponding sites of the onion leaf epidermal wall (FRANKE 1969) and found the same distributional pattern as in the cuticle—a concentration both at the guard cell walls and around stomatal pores, as well as in rows in the middle of the periclinal walls. The second proposal thus appears to be the mechanism by which cuticular penetration occurs.

FRANKE (1968) has further speculated as to the reasons why penetration may be restricted largely to the pathways described above. He considers three possibilities: 1) Binding sites may be determined by specific physico-chemical characteristics of the cuticular constituents, thus making them more permeable. The wall regions directly under such sites consequently become portals for transport; *i.e.*, ectodesmatal localization is predetermined by distribution of permeable cuticular sites. 2) Ectodesmata may result from specific physical or chemical variations inherent in the cell wall. As such they would determine the distribution of cuticular binding sites, the permeability of which may be altered by the action of substances excreted by the protoplast. 3) Localization of both ectodesmata and binding sites may be determined by sites within the plasma membrane, where substances are excreted and perhaps also absorbed.

An experiment by CRISP (1965) would tend to support the first possibility. He found that when the leaves of *Plantago major* were dewaxed by submersion five to ten seconds in chloroform, prior to treatment with the Gilson solution, the previously fan-shaped diffusion patterns of mercury deposition did not occur. Instead, the cuticle was apparently "opened up" sufficiently to allow a more even diffusion pattern as if moving on a front. However, FRANKE (1968) points out that the cuticle is formed after the epidermal walls have been formed, from substances excreted by the epidermal protoplasts. Cutin precursors must consequently penetrate the wall during the course of surface cuticularization. Concomitant with this process, the excretion of aqueous solutions may chemically and physically modify cuticular composition at the emergence sites of such solutions. If such were the case, localization of sites of favored penetrability may be at least partially predetermined by sites of original excretion in the plasma membrane. This would suggest the third proposal as a distinct possibility.

Perhaps the most fundamental problem related to current research on ectodesmata is simply one of interpretation and nomenclature. If these structures cannot be demonstrated by either light or electron microscopy in any way, except by prior fixation with mercuric chloride, the question may very properly be asked: Do these structures really exist? Because of the corrosive properties of mercuric chloride and the varying degrees of dissolution of the cellulose fibrils which it may cause, is it possible we are seeing in some cases (particularly in the electron micrographs) only the results of this dissolution, and not the ectodesmata themselves? Even if this should prove to be the case, the question must still be answered: Why does the sublimate penetrate the cell walls only at specific sites? Could it be *entirely* due to the fact that specific sites in the overlying cuticle are preferentially permeable or capable of binding ions, and that the diffusion pattern of the sublimate into the wall is established only by the location of these sites? This seems unlikely. We would then have to assume that the pathway taken by the sublimate in penetrating the wall is determined at least to some extent by minute anomalies in the dimensions or the orientation of the cellulose fibrils, or perhaps in the chemistry of the intermicellar matrix, anomalies which we have not been able to demonstrate with presently known techniques, outside of fixation with mercuric chloride. The fact that ectodesmata have been demonstrated more frequently in areas which are often subjected to stress or bending, such as guard cells, is suggestive that the looser structure of the microfibrils in such walls allows a preferential penetration of the sublimate. If only slight, but still definite, variations of such structure could be demonstrated within the outer periclinal wall of single epidermal cells, and so done without corrosive fixative, would we be justified in calling such areas of anomalous structure ectodesmata? In view of the microautoradiography experiments showing marked localization of both absorption of various substances and transpiration of tritiated water, and particularly in view of the recent demonstrability of ectodesmata in guard cells of *Ouratea spectabilis without* the presence of a mercuric chloride fixative (ARENS 1968), it would seem we should. It should also be recalled that ectodesmata have been demonstrated by a silver iodide method (SCHNEPF 1959). The question resolves itself, then, not as to whether or not ectodesmata exist, but rather exactly what are the differences in the microstructure and the chemistry of these sites of favored absorption as compared to immediately adjacent sites in the same cell wall?

c) *Leaf hairs*

A hundred years ago HALLIER (1868) established the fact that trichomes and glandular hairs of *Pelargonium* and other plants absorb the large molecules of red cherry juice and indigo-sulfuric acid and emit them again in the epidermal cells of the upper and lower leaf surfaces. From there they were seen to move through the "chlorophyll-free tissue along the veins" and were absorbed by all the elements of the vascular bundles. In the more recent work on foliar absorption the role of trichomes has received relatively little attention. Existing information on this subject has been discussed by MITCHELL *et al.* (1960) and briefly reviewed by several others. Absorption is generally considered to be either cuticular or stomatal. Although entry via the hairs might be considered a type of cuticular absorption (CURRIER and DYBING 1959), the mechanisms involved would seem sufficiently different to warrant separate discussion.

1. Morphology and development.—Numerous types of trichomes are known to exist. For example, METCALFE and CHALK (1950) describe the following six classifications on the leaves of various species of Solanaceae: 1) simple, unicellular, 2) uniseriate of various kinds (widely distributed), 3) variously branched multicellular, 4) various types of tufted and stellate, 5) peltate, and 6) glandular, varying in length and number of cells in the stalks and in size and shape of the heads (widely distributed). A trichome-like structure has been described by BRIAN and CATTLIN (1968) on the leaves of *Chenopodium album*. When young, the leaves are covered with a silvery bloom of spherical globules having an average diameter of about 80 μ. The thin, transparent walls are not entirely bounded by a waxy cuticle and are attached to the leaf by a capillary stalk. These structures, thought to control water relations in the young leaf, collapse and disappear as the leaf matures. Ahmad (1964) considered the trichomes of definite diagnostic value in his cuticular study of 26 solanaceous species. Some were structurally very complex; the stellate trichomes of *Solanum xanthocarpum*, for example, had up to nine rays each. In examining the heteroblastic development of the epidermis and trichomes of the leaves, cotyledons, and other organs of *Crupina vulgaris*, Le Vaillant and Gorenflot (1968) noted eight distinct types of trichomes. Not more than four occurred, however, on any one kind of organ. Petetin (1964), in a taxonomic study of weeds (Compositae and Umbelliferae) from the province of Buenos Aires, found the nature of the trichomes to be of special value in identification of species. The types of hairs, which remained constant under different environmental conditions, could be readily observed with a 12X hand lens.

Structurally, trichomes apparently have thinner walls or less cuticularization near their bases, as suggested by the observation of Haberlandt (1914) that the hairs on *Salvia argentea* leaves trap moisture, which runs to the base of the hair where it penetrates and is subsequently absorbed by the leaf. The lower disk-shaped cells of *Heliotropium luteum* trichomes, intervening between the terminal cells and the embedded basal cells, were observed to have particularly thin walls and were assumed to serve as portals for water penetration. Sifton (1963) noted that elongate leaf hairs of *Ledum groenlandicum* had a transverse orientation of cellulose microfibrils in their primary walls, whereas microfibril orientation was promiscuous over the cell surfaces of capitate hairs. Continuous development and polymerization of cutin occurred over the expanding surface of each hair, as shown by gradually increasing resistance to dissolution by sodium hydroxide solutions. By early May cuticle was prominent on most long hairs, except at their bases. Soon afterward the abscission of hairs on the upper leaf surface occurred, but hairs on the lower surface became more firmly attached by further impregnation of their basal walls with a fatty material that later polymerized to form a cutin-like substance. At that time the walls no longer reacted to ruthenium red, thus suggesting that the pectin had become covered; walls did stain with Sudan IV, indicating cuticularization. A transverse orientation of cellulose microfibrils, as indicated by negative birefringence, was likewise observed in the wall of the basal trichome cell of *Cucurbita pepo* by Fridvalszky (1967). However, it was established that fibrillation in the walls of the other cells slanted at varying directions in the different lamellae, thus resulting in an intersected structure. The structure was apparently related to the fact that the basal cell grows principally in width, whereas the others increase in length.

In ultrastructural investigations of the absorptive hairs of *Tillandsia usneoides* DOLZMANN (1964) noted that the stalk cells contained a heavy cuticle, but the basal "absorption cell," which intervened between the stalk cells and the foot cells in the bottom of an epidermal depression, had little or no cuticle. Within the dome-shaped cells of these absorptive hairs he also described an unidentified "intermediate substance" between the protoplasm and the external cell wall. It was capable of swelling, apparently having the function of water storage, and was differentiated from the protoplast proper by a doubly contoured plasma membrane. The possibility of symplastic transport, from one protoplast to the next, is suggested by the presence of large pits in the transverse walls of *Cucurbita* trichomes (FRIDVALSZKY 1967). On the other hand, intercellular transport of an apoplastic nature may be inhibited in some species. For example, the ultrastructural investigations of SCHNEPF (1968) on mucilage-secreting glandular hairs on the buds of *Rumex alpinus, R. maximus,* and *Rheum rhabarbarum* demonstrated that the outer walls of the stalk cells and the transverse walls between them were completely cutinized. Diffusion along such walls would no doubt be very limited. Numerous species apparently do have ectodesmata in their walls, which should be an aid to penetration. These structures have now been demonstrated by FRANKE and PANIC (1967) between the tip and the base of the cone-shaped trichomes of wheat leaves.

Chemically, trichomes are probably similar to epidermal cells, although there is some indication that differences do exist. MIROSLAVOV (1959) found the content of peroxidase, reduced ascorbic acid, carbohydrate, and tannins to be higher in the trichomes of certain plants than they were in the epidermal cells. He concluded that metabolic activity in the trichomes was particularly high. In vital staining studies of epidermal cell vacuoles with methyl red, ZÖTTL (1960) noted a rich storage of the dye in tannin-containing cells, and a selective staining of guard cells. With *Vicia faba*, however, the glandular hairs showed selective staining and guard cells remained unstained. Investigating trichome fragments of tobacco leaves by means of gas chromatography and other techniques, CHAKRABORTY and WEYBREW (1963) were able to identify 15 of the numerous compounds isolated. Of these, 11 had been previously reported as constituents of whole leaves. Methyl propyl ketone, methyl isobutyl ketone, isooctyl phthalate, and *n*-undecyl acetate were the newly identified constituents of tobacco and may be present only in trichomes.

Although the development of trichomes may not be greatly influenced by environmental conditions, certain inorganic and organic substances can alter normal morphogenesis. BROWN *et al.* (1961) could detect no influence of temperature, air pressure, or light variation upon the development of tobacco leaf trichomes. However, deficiencies of boron, copper, or manganese sometimes resulted in slightly decreased numbers of trichomes reaching maturity. A zinc deficiency caused not only a 25 percent reduction in trichome numbers, but also resulted in the production of monsters, degenerative, and various other aberrant forms. The initiation of unicellular hairs on *Sinapis alba* hypocotyls is known to require light. Stein (1967) has also observed that application of 0.75 μg. of actinomycin D per seedling reduced hair density from 70 to 15/mm². This hair suppression was irreversible, in contrast to the simultaneous retardation of hypocotyl elongation which also occurred. In examining the histological basis for axillary bud inhibition in pea, Sorokin and Thimann (1964) found that young leaves within the buds on isolated stem segments developed two types of trichomes when growth had been inhibited by either IAA or NAA, especially when it had not been completely inhibited. These included 1) short multicellular hairs growing out sporadically from the basal portion of the main inhibited bud, and often on the rudimentary bud, and 2) very long trichomes consisting of three cells each. The shorter ones remained alive longer and were capable of reducing Janus Green B to diethylsafranin. Such trichomes are not normally seen on pea leaves. Their formation was considered to be due to an inhibited supply of water or solutes to the developing bud, perhaps involving imperfect vascular connections.

2. Rôle in foliar absorption.—A thick matting of leaf hairs is known to make foliage difficult to wet; on the other hand, individual hairs may serve as portals of entry for spray solutions. The question is sometimes asked, which of these factors is the more important one, *i.e.*, overall, might leaf hairs be expected to enhance or inhibit penetration of pesticide sprays? Such a question can hardly be answered for plants in general, because of the tremendous variation among species as to the density and structure of their trichomes. Also, leaf maturation, with resultant changes in cutinization of the trichome wall, is doubtless of importance. However, if we can assume that uptake of water or water-soluble solutes bears some relationship to transpiration capacity (e.g., see Linskens *et al.* 1965), the work of Hendrycy (1968) may be pertinent. He cut off the trichomes from the upper surfaces of *Verbascum thapsus* leaves and, seven days later, removed the leaves and placed them in potometers. Leaves without trichomes transpired at a rate about double that of leaves with trichomes intact. Apparently the relative humidity within the thick canopy created by the trichomes resulted in a vapor pressure gradient of sufficient intensity to inhibit transpiration significantly. The greater transpiration in trichome-free leaves resulted in a 23 percent greater uptake of ³²P from the nutrient solution than occurred in normal leaves. The effect on foliar absorption, however, was not investigated. By damaging epidermal hairs of tobacco leaves with a camel's hair brush Volk and McAuliffe (1954) demonstrated a ten-fold increase in absorption of ¹⁵N-urea. In experiments of this nature there is of course at least some possibility of absorption through the broken trichome bases, even though time is allowed for them to heal. In order to eliminate this possibility experiments could perhaps be performed utilizing leaves which are very similar in all respects, except that one should be glabrous and the other pubescent. ORMROD and RENNEY (1968) have suggested *Hypochaeris radicata* and *Taraxacum officinale* as two ideal species for studying the effect of trichomes on retention and penetration. The former has large trichomes and the latter is glabrous, but both apparently have quite similar stomata and cuticular development.

Although the above experiments suggest that trichomes are unimportant as pathways for either transpiration or penetration (since both are greater when they are removed), there is good evidence that they do serve as portals of entry. LINSKENS *et al.* (1965) review information suggesting this to be a principal entryway of aqueous solutions because of reduced resistance. That the minimum resistance occurs in the basal portion of the trichomes, or in the basal cell of multicellular trichomes, seems well established. STRUGGER (1939) reported the fluorescence microscopic observation of "intensive reflection" that the basal membranous portion of the leaf hair exhibited, apparently an area of favored absorption. In further investigations with the fluorescence microscope BUTTERFASS (1956) observed foliar penetration of potassium and sodium fluorescein (uranine), berberine sulfate, oxypyrenetrisulfonate, and rhodamine B via the club-shaped hairs of *Vicia faba* and *Phaseolus coccineus*. Using one-hour absorption periods of the fluorochromes (0.1 percent), the effects of pH and other factors were investigated. Again, maximum penetration was observed at the basal ring of the cuticular hairs from whence movement into the anticlinal walls of the epidermal cells occurred. TIETZ (1954), after studying foliar absorption and translocation of the insecticide demeton, concluded that absorption occurred through the basal cells of hairs and stomatal ridges rather than through the stomatal pores. The relationship of these absorption studies to the previously described observation of SIFTON (1963), in regard to delayed cuticular development of the basal portion of *Ledum* trichomes, is undoubtedly of significance.

Even though substances absorbed by the trichome base move into the epidermal cells, there is apparently some tendency for a polar transport toward the trichome tip. SCHUMACHER (1936) observed such a movement of fluorescein in the leaf hairs of *Cucurbita pepo*. Simultaneously, a vigorous rotation of the protoplasm was visible, but movement of the dye continued independently of it, apparently along the inner border of the cell wall. In a recent treatise on transport of fluorescein in hair cells, SCHUMACHER (1967 b) describes penetration of the basal cells with sodium and potassium fluorescein, and esculin, and further transport from cell to cell within the epidermis. Only the living protoplasm becomes stained, the cell wall and vacuole remaining completely dark. Polar movement of the dye toward the

trichome tip, independent of protoplasmic streaming, was again verified for *C. pepo* and also for *Primula sinensis.* Transport velocity
was similar to that of diffusion; both an increase in dye concentration
and temperature increased velocity. The polarity of movement was
apparently not caused by polarity of the transport mechanism itself,
but rather by the elimination of transported substances by the terminal
cells. A similar polar movement was observed by BENDA (1964) in the
single-celled corolla hairs of *Stapelia grandiflora.* A flow of vacuolar
material occurred when the tip of the hair was cut away and the basal
region (attached to a small amount of corolla tissue) was supplied with
water. The resultant bulk flow was independent of concomitant protoplasmic streaming and did not cause an outflow of cytoplasm from the
cell tip. It was likened to Münch's classical model for the translation of
diffusion into mass flow.

In studies of the foliar penetration of fluorochromes in various
desert plants, HULL (1964 a and 1967 a) observed marked differences
among species with respect to the relative absorption via cuticle, stomata, and trichomes. The absorption by each of these three pathways
is additionally affected by foliar development and, in the case of trichomes, by their location on the leaf. For example, the stellate trichomes
of *Sphaeralcea laxa* are evenly distributed over the leaf surface (HULL
1967 b) but, when near a major vein, they are centered exactly on the
vein. Incandescent microscope illumination fails to distinguish any differences in size or structure between the latter trichomes and those in
intercostal areas. However, if the leaf is irradiated with ultraviolet light
following a two-hour absorption of 0.1 percent acridine orange (plus
one percent of a nonionic surfactant), certain differences become evident. Virtually all of the trichomes in intercostal areas fluoresce a cream
color or very light yellow, whereas a majority of those located on veins
fluoresce dark orange, thus suggesting specific differences in the chemical nature or adsorptive properties of the two types of trichomes. This
difference in trichomes, along with the increased incidence of ectodesmata in cells near veins (FRANKE 1967 a), may explain partially the
enhanced absorption immediately adjacent to veins which has been
widely reported. In evaluating the effect of a nonionic surfactant upon
foliar absorption of auramine 0 over a 90 minute period, we found considerable absorption of the fluorichrome by the trichomes of recently
matured leaves of *Prosopis juliflora,* even without any surfactant present. During the same time no visible quantity of dye penetrated the
stomata or cuticle, in essential agreement with the findings of CURRIER
and DYBING (1959); however, the surfactant-free dye did enter the
stomata and cuticle of old overwintered leaves to some extent, but not
nearly so intensely as it did the trichomes. Stomatal entry in the latter
case may have been partially due to inability of stomata in the old
leaves to close completely.

An interesting pattern of acridine orange absorption via the trichomes of *P. juliflora* has been observed (HULL 1964 a). When leaves were collected one week following budbreak in the spring and treated with this fluorochrome, the trichomes soon emitted their usual brilliant yellow fluorescence. The dye then began to radiate outward from the trichome base into the surrounding epidermal cells. It formed red, yellow, and green fluorescing rings of increasing radii, in much the manner of a circular chromatogram. After a two- to three-hour absorption period, the radii of the outer rings reached 100 μ or more, where it blended into the blue of the background. The same pattern was also exhibited around a few stomata, but the mean radius of spread was of lesser magnitude. When leaves a month more mature were similarly treated the "chromatograms" were sparser and smaller; they would exhibit only one or two, instead of three colors. Several other species were examined for this pattern of absorption, but only on the leaves of *Datura meteloides* was it again observed, and only with acridine orange, of the numerous fluorochromes tested. Leaves of this plant have three types of trichomes, namely 1) long, pointed, 2) shorter, with a slender stalk and globose head, and 3) very short, with a thicker stalk and also a globose head. Only the latter type visibly transported fluorochrome into the epidermis. Since acridine orange emits green fluorescence as the monomer and red as the dimer, the concentric colored circles observed may indeed have represented a true chromatographic separation of the two polymers. The work of HÖLZL and BANCHER (1967) is of interest in this connection. Investigating the vital staining of the inner epidermis of onion scales with different fluorochromes, they noted a strong fluorescent staining of the nucleus but a weaker hue of the cytoplasm following application of acridine orange. This was apparently due to the accumulation of monomeric dye cations in the polar cytoplasmic lipids. The accumulation in lipids was also verified with an artificial system of oil and blood plasma and suggested that the cytoplasmic lipids serve as a vehicle of lipid transport in a mainly aquatic, molecular-disperse phase. HONSELL (1965) has also observed this accumulation of certain vital dyes, depending upon their solution into the lipophilic components of the cytoplasm. The accumulation of anionic dyes was apparently based on their adsorption by hydrophilic components and may be involved in the metabolically dependent staining of cytoplasm by fluorescein. By means of microspectrophotometry, following staining with neutral red DRAWERT and RÜEFFER-BOCK (1965) also examined the onion scale epidermis. The color of the cell sap was apparently not a function of pH, but rather was dependent upon the content of flavonol. Cells with a certain content of this material showed an absorption peak of 560 nm., whereas flavonol-free "empty" cells peaked at 488 nm. Following vital staining of the leaf sheaths of *Agropyron repens*, LUHAN (1963) observed differences in the apparent flavonol content of different epidermal cells. The shorter cells, including guard cells, became "full" much earlier during their development than did long cells and accessory cells. MAERCKER (1965 b), using vital staining, autoradiography, and histochemical techniques, likewise observed differences between the guard cells and the ordinary epidermal cells. Following feeding with ^{14}C-phenylalanine, guard cells showed a greater accumulation of secondary products and a richer content of SH groups. Five enzymes were apparently localized in the guard cells.

Returning to trichomes, in addition to being one of the pathways for entry of pesticides and other chemicals, trichomes are now considered as possible portals for the introduction of bacterial and viral infections. For example, the work of LEWIS and GOODMAN (1965) has suggested that the fire blight bacterium, *Erwinia amylovora*, can enter the astomatous upper surface of apple leaves through trichomes and hydathodes. As has been shown to be the case with dyes and other substances, pathogens apparently penetrate principally at the base of the trichomes. LAYNE (1967) concluded that the foliar trichomes of tomato were particularly favorable sites for infection by *Corynebacterium michiganese*; infection occurred primarily on the upper surface of younger leaves, which had a significantly greater density of hairs than older leaves. Injury sometimes resulted in up to 65 percent of the infections being positively associated with the bases of trichomes. By means of microautoradiography following inoculation of tobacco and *Datura stramonium* leaves with ^{14}C-labeled tobacco mosaic virus, BRANTS (1965) was able to show that β-tracks on the stripping film often occurred in the neighborhood of ectodesmata, especially near trichome bases. Ectodesmata were considered to be likely portals of entry for the virus.

d) *The protoplast*

Assuming entry via the cuticle and periclinal wall of the epidermal cell, a penetrating molecule is next confronted with the plasma membrane. Depending upon the nature of the molecule at the time it reaches this membrane, it may or may not gain entry to the living symplast. Some compounds are excluded at this point and are further transported only within the cell walls, via the apoplast. Others do gain entry to the cytoplasm and are transported primarily therein. Still others are transported to varying extents in both symplast and apoplast. With growth regulators and herbicides, the exast pathway of numerous compounds has been worked out in detail and will be considered in section IV. Certain other individual pesticides doubtless move by specific or preferential pathways in much the same manner. The factors that determine which route a given molecule will follow have not been completely established. They are not entirely related to molecular size, polarity, or solubility characteristics; they may be somewhat dependent upon the affinity certain cellular constituents have for binding molecules of specific molecular configuration. Because of the complexity of both structural and permeability characteristics of the cellular membranes, a detailed consideration is beyond the scope of the present discussion. Both of these aspects have been reviewed recently; consequently, only a few additional findings of interest will be considered. Although here categorized under the epidermis, many of the principles mentioned probably hold for cells of the leaf mesophyll as well. Structure of the plasmatic membranes and penetration through them will be considered separately.

1. Membrane structure.—Recent reviews dealing principally with the ultrastructural and chemical nature of cellular membranes include those of BENSON (1964), BOLIS and PETHICA (1968), BOROVYAGIN (1967), BRANTON (1969), GEL'MAN (1967), HOKIN and HOKIN (1965), LOCKE (1964), LOEWENSTEIN (1966), NORTHCOTE (1968), ROBERTSON (1967), and ROGERS and PERKINS (1968). The plasma membrane and tonoplast have been considered to consist of a central, bimolecular leaflet of phospholipid bounded on either side by proteins and possibly carbohydrates. FRANKE (1967) and several of the above reviewers discuss research which suggests a closely-packed globular substructure in these membranes. FREY-WYSSLING and MÜHLETHALER (1965) describe this as a submicroscopic mosaic consisting of areas of larger globular particles between areas of smaller such particles. Work with negatively stained preparations, reported by GLAUERT (1968), led to the suggestion of a new model for membrane lipids, namely that some of the lipids may be organized in small globular micelles, while the remainder are in bimolecular leaflets. It was considered that changes in proportion of the lipid in the two configurations may account for some of the specific physiological properties which different membranes possess.

By means of the freeze-etching technique, BRANTON (1966) was able to split cellular membranes and expose the internal faces. Small particles of 75 to 200 A diameter found on these faces were interpreted as subunits of the membrane. Their numerical and stuctural relationships were a function of the type of membrane. He also considered the membrane apparently to be organized in part as an extended bilayer and in part as globular subunits. The freeze-etch technique was also used by STAEHELIN (1968) in ultrastructural studies of both artificial and natural membranes. He observed particles in all of the biological membranes examined. With the exception of the chloroplast and mitochondria membranes, these particles could be removed in such a way as to demonstrate an underlying central layer with fracturing characteristics similar to bimolecular lipid leaflets. It was considered that the different freeze-etch images of the membranes could be caused by the fracture planes running at different levels through or along the upperlying halves of the membranes. MARCHANT and ROBARDS (1968) make the suggestion that all membranous or vesicular structures associated with the plasma membrane be classified under the general term "para-mural body," regardless of their origin. Such bodies would be subdivided into two classes, according to their derivation, thus "lomasomes" formed from cytoplasmic membranes and "plasmalemmasomes" formed entirely from the plasma membrane. During an ultrastructural study of moss (*Mnium cuspidatum*) leaves of varying ages, LÜTTGE and KRAPF (1968) likewise found that leaf cells of the younger shoots had numerous particles or vesicles within the tonoplast, the presence of which could be directly related to previously reported differences in ion absorption. From a functional standpoint, the report of PARDEE (1968) characterizes the role of proteins in the membrane transport system. He reviews recent evidence that the membrane consists of a protein-lipid mosaic, with proteins extending through the membrane in some places, providing specific doors for transport processes. Such proteins apparently constitute "recognition sites"; they are the only molecules having the observed degree of specificity to discriminate between possible substrates, *i.e.*, the smaller molecules being transported. Such transport proteins are being isolated and characterized.

Although apparently single-layered membranes have been reported, both plasma membrane and tonoplast usually appear in cross section as two symmetrical thin dark lines separated by a light region, with a total width generally ranging from 70 to 100 A. This basic structure, for example, has been observed by LEDBETTER (1962) in the filament cells of *Triticum aestivum* and by SCHNEPF (1960) in the gland cells of *Drosophyllum lusitanicum*. MERCER (1960) reviews evidence of this frequently double, and apparently homogeneous, structure of the plasma membrane. However, GRUN (1963) has observed that the plasma membrane in the cortical cells of *Solanum* root tips is symmetrical and about 75 A wide, whereas the tonoplast is either symmetrical or asymmetrical. If the latter, the wider of the opaque layers was the inner one, on the vacuole side of the tonoplast. Symmetrical tonoplasts were about 75 A wide, having inner and outer opaque layers of 25 A. The asymmetric membranes were similar in structure except that the inner layer varied in width from 25 to 250 A. Tonoplasts were the only asymmetric membranes seen.

The structural visualization of cellular membranes appears to be markedly influenced by method of fixation. Both potassium permanganate and osmium tetroxide have been used for specimen fixation in some of the works described. In a review on plant membrane lipids, BENSON (1964) points out that these two compounds may reveal images which are not at all identical. Permanganate reacts readily with carbohydrate, whereas osmium tetroxide does not. Glycolipids thus stain more effectively with the former fixative than with osmium. It has been suggested by SJÖSTRAND (1962) that the different structures visualized by the two fixatives may be because the plasma membrane is asymmetric, with a thicker protein layer on its cytoplasmic side and a thinner protein or polysaccharide layer as the most peripheral constituent.

The latter layer would not show up in osmium-fixed material, but only in tissue fixed with permanganate. These variable images, resulting from different fixatives, are not confined to membranes. Esau (1965 b), for example, has observed some striking variations in the plastids of *Beta*. A prominent ringlike portion of the plastid, apparently a proteinaceous inclusion, is preserved by osmic acid but is eliminated in fixations with potassium permanganate. The external sculpturing of the plastid likewise is not preserved by permanganate. She discusses the relative advantages of these fixatives, and also of glutaraldehyde.

2. **Penetration of and binding within plasmatic membranes.**—Membrane penetration, in addition to being discussed in some of the reviews and monographs mentioned in the previous section on structure, has also been considered by Bolis *et al.* (1967), Brian (1964), Collander (1959), Dainty (1964), Hendricks (1964), Jennings (1963), Laties (1969), Lüttge (1968), Quastel (1965), Stein (1967), and Troshin (1966). Detailed information on membrane carriers, the kinetics of transport and other pertinent information may be found in the above reports. Generally speaking, a membrane showing selective permeability characteristics, and which is situated between two aqueous phases, will present three distinct types of resistance to a penetrating molecule: 1) resistance in passing from one phase into the membrane, 2) diffusion resistance within the membrane, and 3) resistance in passing from the membrane into the second aqueous medium. Brian (1964) reviews evidence indicating that molecules which form three or more hydrogen bonds with water are limited by stage one. Nonpolar molecules having seven or more -CH_2- groups are apparently limited by stage three, whereas those with oil/water partition coefficients of 0.1 to 0.005 are limited principally at stage two.

If a molecule is to be transported within the symplast, it must of course penetrate the plasma membrane following its diffusion through cuticle and cell wall. Although additional penetration of the tonoplast is not essential for continued movement in the symplast, the permeability characteristics of this membrane are involved in establishing the ultimate destination of a penetrating molecule. Differences in the permeability of plasma membrane and tonoplast to uranin have been clearly demonstrated by Bolhàr-Nordenkampf (1966) in the inner epidermal cells of *Allium cepa* scales. The fluorochrome readily penetrated the semi-permeable plasma membrane when in the undissociated form at pH 3.1. It was, however, impeded somewhat in penetrating the tonoplast, apparently because of the storage competition of the vacuolar plasma, although penetration into the vacuoles of live cells could be induced with an ammonium carbonate solution. After replacement of the solution with water, the uranin was again shifted from vacuole to cytoplasm, but the vacuolar loading became irreversible if the stained slices were left several hours, very likely because of binding to some metabolic product of plasmatic origin. In this connection

BANCHER *et al.* (1968), also using *Allium* epidermis, have now determined that uranin is chemically changed during transport from cytoplasm to vacuole. Chromatographic and spectrophotometric analyses indicated apparent esterification of the fluorochrome by the action of acetyl-coenzyme A following penetration of the tonoplast. Since the physical and chemical properties of certain pesticides are not greatly different from those of uranin, it seems plausible that their relative penetrability characteristics might be similar in some respects to those of uranin.

As pointed out in the "Introduction," this review pertains principally to absorption in the intact plant, or at least the entire leaf. Uptake within individual cells can sometimes be directly studied in the smaller algae, or perhaps in the thin scale leaf of the onion. However, to eliminate the cuticular barrier in most of the higher plants it is generally necessary to use leaf disks or slices, or root segments, or even isolated cells. Investigating the uptake of oxalate and various monovalent cations by the mesophyll cells of *Atriplex* leaves, OSMOND (1968) observed that disks did not permit equilibration of interior cells with the bathing solution. However, slices 500 μ wide did allow rapid access of electrolyte to all cells. SMITH and EPSTEIN (1964), in their studies on absorption of rubidium ions by maize leaves, found that by using slices 300 μ wide, absorption approximated that of excised barley roots more closely than when leaf disks were used. Behavior of this nature is apparently due to the relatively greater surface exposure of parenchyma cells. RAINS (1968), in studies on the kinetics and energetics of light-enhanced absorption by maize leaves, likewise observed that with narrow leaf slices absorption of potassium ions occurred through the cut edges and not through the surface of the leaf lamina. In the work which follows, the assumption is made that uptake by an intact mesophyll cell within the leaf will follow essentially the same patterns defined by means of narrow cell strips, stem segments, or isolated cells.

It is apparent that the relative permeability of plasma membrane and tonoplast to organic substances is not unlike that which exists for certain inorganic ions. For example, the flux measurements and calculated permeability coefficients for sodium, potassium, and chloride ions in *Nitellopsis* reported by LATIES (1964) make it evident that the plasma membrane and tonoplast differ markedly in their permeability characteristics toward these ions.

After studying the uptake of chloride ions in both nonvacuolate maize root tips and vacuolate subapical root sections, TORII and LATIES (1966) concluded that two distinct systems exist: one of high-affinity which mediates ion passage through the plasma membrane, and one of low-affinity which implements transport through the tonoplast. The first system operates within the range of 0.02 and 0.5 mM, and is apparently metabolically controlled and carrier-mediated. At higher concentrations from one to 50 mM, this system breaks down, in that further uptake via the plasma membrane becomes nonspecific and nonmeta-bolic—essentially diffuse permeation. In the vacuolate tissues the second system continued to be operative under metabolic control, through the higher concentration ranges. The uptake of rubidium ions followed a similar pattern, although they were differently affected by the presence of certain counterions. This dual mechanism of ion absorption was corroborated by OSMOND and LATIES (1968) in disks of tissue cut from *Beta vulgaris*. They also observed that when the cytoplasm was prefilled with the analogous unlabeled salt, a diminution of subsequent isotope uptake therein occurred only in the concentration range of system one. This and other data suggested that the cytoplasm acts as a mixing chamber in conjunction with absorption via the two systems described. By determining ion fluxes at the plasma membrane and tonoplast as a function of concentrations in both cytoplasmic and vacuolar phases, and also the external medium, LÜTTGE and BAUER (1968) were able to confirm the above hypothesis in both maize and the moss *Mnium cuspidatum*. Noting the changes in amounts of internal ions with changes in external concentration, they also deduced that the plasma membrane loses its function as a barrier at concentrations above one mM. Additional confirmation of ion uptake of this nature into the cytoplasm of *M. cuspidatum* leaf cells was found by LÄUCHLI and LÜTTGE (1968). By the use of $^{35}SO_4^{--}$ for microautoradiography and potassium dihydrogen phosphate for electron probe X-ray microanalysis, they demonstrated a greater concentration of label/unit of volume in the cytoplasm than in the vacuoles. The pattern of cytoplasmic filling with time, its relationship with external concentration, and other kinetic features were compatible with those of the previously-described system one.

A pattern of borate absorption similar to that described above, but modified in some respects, has been observed by BOWEN (1968) in excised sugarcane leaves. Phase I in this process consisted of a rapid and reversible influx into leaf mesophyll cells. This was followed by phase II, a slower and irreversible accumulatory phase. The latter phase was represented by three distinct absorption mechanisms, each dependent upon the external concentration. The highly specific first two mechanisms transporated borate across the initial barrier into the cytoplasm, and the third reaction carried the ions across the tonoplast. Certain ionic antagonisms and actions of specific metabolic inhibitors suggested a linkage of the first two mechanisms to respiratory electron transport and of mechanism number three to oxidative phosphorylation.

In contrast to the concept of system one being operative in the plasma membrane and system two in the tonoplast, WELCH and EPSTEIN (1968 and 1969) have presented evidence that even at high concentrations, where system two is operative, the plasma membrane does not become permeable to chlorine or certain monovalent alkali cations. The two systems apparently operated in parallel, and both within the plasma membrane. Thus, when barley roots were plunged in a chloride solution the transport mechanisms began operation at a rate dependent upon prevailing conditions and maintained this initial rate at least an hour. This occurred regardless of whether the ionic concentration was in the low range, where only system one is operative, or was higher where the type two system also becomes active. The steady absorptive rates were established so quickly that it made no difference whether they were measured over a one-, a ten-, or a 60-minute period. Participation of the tonoplast seemed impossible when a steady absorptive rate was achieved so rapidly.

Various internal and external factors are apparently capable of altering the primary state of the plasma membrane, in which it is essentially impermeable to certain ions, and thus enhancing its permeability. For example, using *Vallisneria* leaves, ARISZ (1964) has shown that uptake of asparagine is an active process, displaying essentially the same features as salt uptake. In its primary state the plasma membrane was impermeable to asparagine ions, which are presumably dipolar; but in the secondary state the permeability was increased, following the antagonistic effect of potassium and calcium ions. The nonpermeability of the membrane could be preserved, however, by the maintenance of a balanced solution.

The effects of different growth substances on metabolic uptake of IAA in maize mesocotyl segments and of both IAA and NAA in oat coleoptile and bean hypocotyl segments has been studied by SABNIS and AUDUS (1967 a and b). They observed that, although naptalam stimulated IAA uptake in maize, 2,4-D was without effect. Other substances including NMSP, TIBA, DNP, ioxynil, and bromoxynil gave slight to marked inhibition of IAA and NAA uptake, depending upon the regulator, its concentration, and the plant species involved. The metabolic uptake of ^{14}C-labeled 6-BAP and adenine by barley leaf segments was followed by DULAEVA *et al.* (1967). After saturation was attained, the uptake of both compounds continued throughout the three-day period studied, thus suggesting that the compounds were continuously consumed during metabolism of the leaf cells. The intensity of 6-BAP uptake was much smaller than that of adenine, indicating that the 6-BAP was metabolically consumed to a much lesser extent than adenine. With dicamba, MAGALHAES and ASHTON (1969) demonstrated that permeability of cellular membranes of *Cyperus rotundus* leaves was significantly decreased following treatment with this herbicide at 10^{-2} or $10^{-3}M$. Such membrane disruption may be related to the mode of action of this compound.

Some interesting effects of sodium fluoride on absorption of ^{32}P from $NaH_2{}^{32}PO_4$ by leaf disks of tobacco have been reported by PENOT (1967). When included in the phosphate medium at 10^{-1} to $10^{-3}M$, sodium fluoride inhibited absorption of the ^{32}P. However, when the disks were pretreated with the fluoride subsequent absorption was stimulated. The stimulation was operative only with mature leaves and its relationship to the effect of light suggested its dependence on cellular metabolism. The specific effect of red light upon ^{14}C and ^{32}P uptake by etiolated maize leaf sections has been studied by STEINER et al. (1968). A brief exposure to such light, given before placing the sections on different substrates, enhanced ^{14}C uptake and its incorporation into various metabolites as compared to the dark controls. Previous work had shown that when the sections were floated on the substrate before irradiation, the effect was in the opposite direction. However, in both types of response, the red light resulted in increased hexose-monophosphate turnover and ^{14}C accumulation in the cell wall polysaccharides. Thus, the regulatory influence of the hexose pool size, as well as the phytochrome responses, was apparently influential in this action.

Important contributions on the uptake of chlorinated phenoxyacetic and benzoic acids by stem tissues of various plant species continue to be reported from Oxford. Of numerous ^{14}C-phenoxyacetic acids SAUNDERS et al. (1965 a) found that the highly active 2,4-D and 2,4,5-T were initially taken up by etiolated stem segments of Gossypium hirsutum and Pisum sativum at a high rate, but that the rate decreased progressively with time and could end in a phase of net loss. In contrast, the weak auxins 2-CPA and POA were accumulated in an essentially linear fashion. Uptake by several gramineous species was different, in that uptake of all compounds was cumulative, even though the final concentrations achieved for 2,4-D and 2,4,5-T were less than for 2-CPA and POA. It is of interest that the two dicots were susceptible to herbicidal concentrations of 2,4-D, whereas the grasses were all resistant. Also significant was the relationship of chemical structure with both pattern of uptake and physiological activity. This relationship was further examined by Saunders et al. (1965 b) in stem tissue of Gossypium. They observed that pretreatment of stem segments with buffer prior to ^{14}C-2,4-D treatment resulted in an initial depression of uptake and a failure of the above-described net loss to occur. In contrast to the buffer-treated segments, uptake by fresh segments had a high Q_{10} and was markedly depressed by both 2,4,5-T and IAA, which suggests that the net loss may result from release of material accumulated by a specific mechanism which, in time, becomes inoperative. Continuing work with the same species, SAUNDERS et al. (1966) demonstrated that uptake of the weak auxin, POA, was inhibited by low temperature, anaerobiosis, 2,4-DNP, and iodoacetate. It was concluded that POA accumulation involves its metabolic conversion to products which do not readily diffuse out into the external medium. Uptake of 2,4,5-T on the other hand was less sensitive to temperature, anaerobiosos, and the above inhibitors. Its net loss from the tissue during the second phase of uptake was balanced by the release of a like amount of 2,4,5-T to the medium. No radioactive metabolites of 2,4,5-T were detected. This and other information suggested that 2,4,5-T uptake involves reversible accumulation by a mechanism whose efficiency decreases with time, the most likely systems being either a metabolically linked mechanism for active transport across a membrane or reversible adsorption on specific binding sites. JENNER et al. (1968 a) subsequently observed in Avena mesocotyl segments the above-described type of second-phase net loss of 2,4-D to the bathing solution, which could be overcome by pretreatment with buffer. Such pretreatment enhanced absorption of both 2,4-D and POA, and resulted in a curvilinear uptake of the former compound but a linear uptake of the latter. If the segments were transferred to a buffer after uptake of the labeled 2,4-D, a small portion of the ^{14}C, the "mobile fraction," was released, the amount released being essentially constant and largely independent of period (and quantity) of uptake. The "residual fraction" of non-mobile ^{14}C, however, rose progressively with accumulation. Both 2,4-D and POA uptake were accompanied by formation of residual labeled metabolites within the tissue, the relative composition of this fraction apparently being interrelated with both accumulation and metabolic conversion. Additionally, JENNER et al. (1968 b) suggested that variation among compounds as to their relative rates of accumulation may be a function of the stability of their conjugated derivatives, and that the facility of conversion may be a factor in determining physiological activity.

Some unique aspects of uptake and binding of 2,3,6-TBA by Avena mesocotyl segments have been reported by VENIS and BLACKMAN (1966). As with the phenoxy compounds described above, 2,3,6-TBA also demonstrated an initially positive uptake which subsequently became negative. The latter phase of net loss, however, could be prevented by streptomycin or Synthalin®, whereas enhanced accumulation was induced with CTAB, a cationic detergent. The authors postulated that initial accumulation is governed by an unstable accumulatory process (Type one) which involves adsorption by some cell membrane system through an interaction between the carboxyl anion of the TBA molecule and the quaternary ammonium group of the choline moiety of a-lecithin. This type of binding is destroyed following hydrolysis of lecithin by phospholipase-D, whereas cationic nitrogen compounds maintain positive uptake by competing with the choline quaternary ammonium group of a-lecithin for the anionic site of phospholipase-D. Additional experiments on the uptake of alkyl pyridinium compounds and on their adsorption to lecithin in vitro also suggested that quaternary ammonium compounds, such as CTAB, act largely by providing artificial Type one sites. That herbicides of the quaternary ammonium type can be highly damaging to the plasma membrane has been demonstrated by the ultrastructural studies of BAUR et al. (1968). They found that treatment of Prosopis glandulosa leaflets with 0.01 M paraquat for five minutes resulted in disintegration of the plasma membranes in 39 percent of the mesophyll cells. Disruption of the chloroplast membranes required 40 minutes.

Various techniques have been used in determining the mechanisms involved in the binding of pesticides and various other compounds. In continuing studies of adsorption of plant growth regulators to monomolecular layers of oat squashes, BRIAN (1967 a) has utilized a new technique involving the competition of two organic acids for a limited number of sites on the monolayer. A standard acid was selected for its ability to induce a large surface potential change when adsorbed on the oat monolayer, 2,4,5- and 3,4,5-trichlorophenylacetic acids being most effective. One of these acids was applied as a mixture with the acid to be assayed for adsorbability, by injection under the monolayer. The acids thus would compete for adsorption sites on the surface layer. Benzoic and phenoxyacetic acids were not adsorbed, but phenylacetic acid was weakly adsorbed. The second ring in naphthalene and naphthoxyacetic acids greatly enhanced adsorption. Substitution of the 2- and 6-positions in the phenyl and phenoxyacetic acids resulted in low adsorption, but 2,6-disubstituted phenoxybutyric and benzoic acids were more highly adsorbed. With the exception of chlorine in the 3-position of the aromatic ring, no general relationship existed between adsorption of the various acids and their activity. The possible role of the amido group as an adsorption mechanism was investigated by WARD and UPCHURCH (1965) in an artificial system. Adsorption of some 52 structurally related N-phenylcarbamates, acetanilides, and anilines was quantitatively evaluated on Nylon 66 [containing an ordered arrangement of amido (-NHCO-) groups similar to those found in proteins], cellulose triacetate, and cellulose. Nylon and cellulose acetate adsorbed all chemicals to varying extents, whereas cellulose adsorbed none. Desorption studies showed that adsorption occurred through hydrogen bond formation. The inverse relationship which occurred between solubility and adsorption accounted for 60 percent of the total variation in adsorption. The preferred adsorption mechanism of the amido compounds from an aqueous solution was via the adsorbate's imino hydrogen and the adsorbent's carboxyl oxygen. If neither of these was available, alternative binding sites were utilized.

A few studies have been carried out by means of enzymatically isolated cells. YAMADA et al. (1965) examined the effect of urea on absorption of rubidium and chloride by palisade and spongy parenchyma cells enzymatically separated from tobacco leaves. Urea enhanced rubidium absorption, although it had no effect on chloride uptake. This is in contrast to the authors' previously-mentioned studies with isolated tomato fruit cuticle, in which a urea-enhanced penetration of both ions was observed. KANNAN and WITTWER (1967 a), also using enzymatically isolated cells of tobacco leaves, found iron absorption to be temperature sensitive, and to be increased by light and succinate, which apparently served as energy sources. This, and the activation energies found, suggested an energy-mediated absorption. They also observed that ^{59}Fe uptake by isolated cells from chlorotic and healthy leaves of two cultivars of soybeans, differing in susceptibility to chlorosis, showed marked differences. The results indicated a possible genetic control of nutrient requirement, and a rate limiting cellular absorption. JACOBY and DAGAN (1967) followed the net sodium fluxes into thin bean leaf sections, and found it to be 65 times higher than into enzymatically isolated mesophyll cells of the same leaves. The isolated cells did not stain with neutral red, whereas the dye did accumulate in similar cells in small tissue segments which had not been separated by enzymatic treatment. The authors concluded that the plasmatic membranes of isolated bean leaf cells are damaged, and hence unfit for studies of ion uptake. A related point of interest involves the possible damage to plasmatic membranes which may result from various types of radiation and from pathogenic organisms. In studying the plasmolysis and permeability of cells of the upper epidermis of *Allium cepa* scales, STADELMANN and WATTENDORFF (1966) observed that a-irradiated tissue placed in a perfusion chamber showed incipient plasmolysis with $1M$ glucose a little more rapidly than the non-irradiated controls. Deplasmolysis in a subsequently applied $1M$ urea solution, however, demonstrated no significant difference in permeability constants for this compound in the irradiated and control samples. BURKOWICZ and GOODMAN (1969) found that when apple leaves became infiltrated with a virulent or avirulent strain of *Erwinia amylovora*, the causal bacterium of fire blight, an increased permeability of cellular membranes was initiated, followed by tissue necrosis. Inoculation concentration and leaf age were principal factors determining rate and extent of permeability alterations. Considerable amounts of K, Na, Ca, P, and Fe could be detected in tissue leachates within 12 hours of inoculation.

IV. Mesophyll and vascular tissue

a) *Patterns of movement within the leaf*

Lateral and longitudinal movement of substances within the lamina of the foliage leaf probably occurs by at least three mechanisms: (1) intercellularly in the mesophyll via the interconnected living symplasm, (2) intercellularly in the mesophyll, but via the apoplast, *i.e.*, within the cell wall or along the interface between cell wall and intercellular spaces, and (3) within the conductive tissues of the veins either by symplastic movement in the phloem or apoplastic movement in the xylem. Combinations of the above types of movements are of course possible. Petroleum spray oils, after either cuticular or stomatal penetration of the leaf, move principally along the intercellular spaces as described in (2) above. ROHRBAUGH (1934) noted this type of movement in citrus leaves and, also, that oil could remain in such a position for two years or more without apparent injury to the leaf. The oil tended to concentrate at the leaf margin and along the midrib, suggesting greater absorption in these regions. Although most substances, both oil and water soluble, do tend to penetrate more readily and be transported out of the leaf to a greater extent if placed directly over the midvein [e.g., see BRIQUET et al. (1968) relating to ^{14}C-2,4-DB movement in bean], this is apparently not always the case. FOY (1962), following the movement of ^{14}C-dalapon by autoradiography, noted that entry was more apparent into the maize leaf when application was off the midvein than when directly over it. This may have been partially due to the lignified bundle sheaths about the major veins. In spite of this greater penetration at the intercostal areas, buildup to a high concentration occurred only when application was made over such areas, as compared to over the midvein, thus suggesting limited intercellular movement. Studying absorption of ^{14}C-leucine in apple leaves, KAMIMURA and GOODMAN (1964 b) likewise observed greater absorption at the intercostal positions on the lamina. They could not invoke translocation away from such areas. Greater absorption was also noted at apical positions on the lamina than at basal positions.

The exact velocity at which different compounds penetrate from leaf surface to the finer veins within has not been studied to the extent that has velocity of movement within the conductive tissues of stems, coleoptiles, etc. There are some approximations, however. By application of a droplet containing 2,4-D to the upper surface of a primary bean leaf, and subsequently punching out the treated area at various intervals following treatment, DAY (1952) determined (by subsequent epicotyl epinasty and prior knowledge of velocity in the petiole) that the compound moved from leaf surface to the phloem within at a rate of 25 to 35 μ/hour. Also using bean, LITTLE and BLACKMAN (1963) studied movement of ^{14}C-labeled 2,4-D and 2,4,5-T both by epinastic response and by autoradiography. They established that transport of either phenoxy compound from the leaf surface to the underlying vascular bundles takes 79 minutes, which gives an average velocity only slightly lower than that determined by DAY. Movement of fluorochromes within walls of the mesophyll cells has been shown by SCHLAFKE (1958) to be slower than diffusion in gelatin, and to be unaffected by inhibitors. The velocity of such movement in Elodea densa ranged from 4.2 to 7.5 μ/hour. Movement was considerably more rapid in Stellaria media, and was dependent both on the fluorochrome used and whether transport occurred in the cell walls or along the midvein. Rates up to about 30 cm./hour were observed in the latter.

The relative importance of vascular bundles and parenchyma on transport within the leaf has been discussed by HELDER (1967), who considered much of the pioneering work of ARISZ and his colleagues.

For example, when the parenchyma was excised from *Vallisneria spiralis* leaf segments in such a way as to leave the central bundle intact, or when the central and two lateral bundles were excised while leaving two parenchyma bridges intact over distances of four, eight, or 16 mm., transport from the "absorbing part" through the bridges to the "free part" could be measured. Transport of chloride across a four mm. bundle bridge was about equivalent to that in the intact controls, and was only slightly reduced with the longer bridges. Movement through the parenchyma bridges was somewhat less than through the bundle bridges, but was still very significant. Such transport was, however, more sensitive to bridge length than was transport through the bundle bridges. With asparagine transport, some experiments showed an equivalent facility of movement through both types of bridges.

As to movement of herbicides within the leaf, the extensive studies of CRAFTS and his colleagues (BAYER and YAMAGUCHI 1965; CRAFTS 1964, 1966, and 1967; CRAFTS and YAMAGUCHI 1964; LEONARD *et al.* 1966; and PICKERING 1965) have clearly demonstrated distinct mechanisms of transport for different compounds. Autoradiography has shown, for example; that the phenoxy compounds are transported primarily via the symplast; the ureas, phenylureas, and triazines via the apoplast; and amitrole, dalapon, dicamba, and MH via various combinations of these two systems. CHANG and VANDEN BORN (1968) found ^{14}C-dicamba to move in both phloem and xylem and to accumulate in young, growing leaves following either foliar or root uptake. Analysis of the treated leaf after 54 days showed 63.1 percent of the recovered activity to be still in the form of unaltered dicamba. In some studies, e.g. PICKERING (1965), histoautoradiography techniques have demonstrated that although 2,4-D enters the symplast and becomes concentrated in the protoplasm, it may also leak into the vacuoles and the surrounding apoplast. Although present in both the phloem and the xylem of the treated area, translocation was observed only in the phloem.

KREIDEMANN (1967), following the uptake of ^{14}C-sucrose into cotyledons of germinating castor bean seedlings by microautoradiography, found most activity to occur in the vicinity of cell walls and in intercellular spaces. This picture suggests a diffusion pathway via the apoplast as the predominant route of entry; however, it is possible that a portion of the activity represented ^{14}C (not necessarily sucrose) which originally occurred in the cytoplasm, but which was subsequently deposited against the cell walls during freeze drying. ZIEGLER and LÜTTGE (1967) also used microautoradiography to study the movement of ^{36}Cl in the *Limonium vulgare* leaf, and localized the silver precipitation by electron microscopy. The pattern of accumulation in chloroplasts, nuclei, vacuoles, in the cell walls of guard cells, and on the cuticular surface, was suggestive of both an apoplastic and symplastic transport of this ion. The dependence of symplastic movement upon metabolic activity has been rather clearly demonstrated by the autoradiographic studies of YAMAGUCHI (1965) on several plant species. Following spot application of ^{14}C-2,4-D to the leaf surface, he found that metabolic absorption and retention was reduced by DNP. This reduction resulted in a concomitant reduced phloem mobility but enhanced apoplastic mobility. Similar results were also obtained by starvation and anoxia.

CRAFTS (1967) has emphasized the difference in the mechanisms involving movement of exogenous solutes within the vascular system and movement of endogenous substances through relatively undifferentiated tissues. Polarity of movement is of course one of the factors involved. Within the lamina, where transport may be dependent upon either one or both of these systems, some complex interrelationships exist with respect to polarity. Studying the transport of ^{32}P-labeled Rogor®, a systemic insecticide, DE PIETRI-TONELLI (1965) observed that this compound (or a labeled metabolite) moved principally in an acropetal direction following application to bean or lemon leaves. WHITESELL and THOMPSON (1968) demonstrated a similar acropetal transport of Azodrin®, another systemic insecticide, when it was applied to maize leaves. Following uptake it apparently no longer remained in the form of free Azodrin®, and its systemic activity was primarily limited to the leaf to which it was applied. Acropetal movement of this nature was noted in the case of two s-triazines by LÓPEZ AROCHA and RINCÓN (1966) when applied in a lanolin ring on the basal portion of French bean leaves. Rates ranging from 0 to 250 μg./plant gave varying degrees of necrosis within and beyond the ring, ametryne being more active in this respect than simazine. Wedge-shaped necrotic areas developing toward the leaf tip, and a lack of evidence for basipetal movement, suggested essentially apoplastic transport.

The translocation pattern of certain inorganic substances within the leaf is also suggestive of an apoplastic movement. JACOBSON (1966) found atmospheric fluorides and hydrogen fluoride to be absorbed by the entire leaf surface and to be then translocated to the tips and margins. These highly soluble compounds appear to move with the transpiration stream. The leaf cells are apparently incapable of removing and binding fluoride. When applied to the leaf surface of oat plants, RINGOET et al. (1967 and 1968) observed that ^{45}Ca, after slow absorption, is also transported predominantly in an acropetal direction. This movement, apparently in the xylem, occurred at low concentrations. Basipetal movement in the phloem was also observed, but only at higher concentrations of applied solution ($>0.02M$ calcium chloride). The slow and limited redistribution of ^{45}Ca was probably not due to its inability to move in the phloem, but rather to the great accumulation and absorption capacity which the various leaf tissues have for this ion.

In contrast to the essentially apoplastic movements described above, other materials seem to move symplastically within the leaf blade, and largely in a basipetal direction. For example, LEVI (1968) found that ^{22}Na applied to the tip of one primary bean leaf was absorbed and moved in such a manner, even in the dark. Autoradiography showed the ion to be present largely in the vascular system of the lamina, with little or no accumulation in the mesophyll. Isotopes of ^{32}P, ^{43}K, ^{86}Rb, and ^{134}Cs moved in a somewhat similar manner, whereas other cations did not. THOMPSON (1968) also observed a distinct basipetal polarity of both ^{14}C-2,4-D and ^{14}C-IAA transport in isolated peanut leaves, with 2,4-D being the more polar of the two regulators. NAKATA and LEOPOLD (1967) have demonstrated preferential translocation of radioactivity from labeled sucrose, IAA, and phosphate toward the base and petiole of isolated bean leaves, suggesting a natural mobilization gradient down the blade and petiole. Local application of a kinin (BA) three days previously diverted the movement of the sucrose label, and to some extent the phosphate label, by apparent establishment of new mobilization centers. IAA was not mobilized by BA. The data suggest that the unloading of solutes from the phloem at either natural or artificially created mobilization centers may determine both direction and intensity of transport.

Some compounds appear to move with great facility within leaves. One such compound, iodoacetic acid, has been studied for its ability to enhance abscission of citrus fruits, or at least to minimize the tenacity with which fruits are held, thus making possible more efficient picking with mechanical harvesters. This property has been demonstrated on early and mid-season orange varieties, but not on the late "Valencia". TALTON et al. (1967) found by autoradiography following spot treatment of leaves with ^{14}C-iodoacetic acid, that absorption and extensive movement throughout the leaves of both early and late varieties occurred within two hours. Likewise, both varieties were equally capable of decarboxylating the iodoacetic acid. Their differential response, therefore, must be related to some factor other than the leaf transport and decarboxylation mechanisms. Great variation does exist in the extent to which various growth regulators and other substances become bound within the leaf; some are effectively altered during transport within the lamina. Thus, MITCHELL (1963) reported that when ^{14}C-NAAm was quantitatively applied along the outer edge of bean leaves, two days later it could not be detected as such in the main leaf veins, the petioles, or the stems. These parts did contain, however, an initial reaction product which could be chromatographically differentiated from NAAm. Two additional metabolites were detected in the petioles, and the stems contained still a fourth compound which differed from the parent lactamide and all the other reaction products detected. The study clearly indicates how movement of the original intact molecule may be limited, but different parts of the plant are capable of altering it to metabolites which are capable of being transported in various tissues.

Evidence pertaining to the movement of cytokinins and gibberellins is conflicting. Some work suggests that these regulators, when externally applied, are more mobile than endogenous substances. This, of course, depends upon the method of application and other factors. The movement of these compounds has been reviewed by McCready (1966). According to Goldsmith (1968), the movement of kinins differs from auxins in being neither consistently nor immediately polar.

One factor which could conceivably affect transport of pesticides within the leaf is the development of callose, either in the sieve elements of the veins or in the parenchyma. Webster and Currier (1968) demonstrated that heating cotton cotyledon petioles at 45° C. for 15 minutes increased phloem callose and decreased lateral movement of ^{14}C-assimilates. If the petioles were heated immediately after a 16-hour dark period, little callose formed and lateral movement of assimilates was less inhibited, thus reaffirming the inhibitory influence of callose on translocation. A similar inhibition of basipetal transport by heat-induced callose has also been demonstrated in the cotton hypocotyl by McNairn and Currier (1968). As with the petiole, callose deposition on the sieve plates diminished with time, being noticeably less after six hours and virtually normal within two days. Growth measurements, plasmolytic tests, vital staining and visual observations revealed no evidence of injury in the heated plants. Additional factors influencing callose development (Currier and Eschrich 1964, Currier and Shih 1968, Currier and Webster 1964) include leaf detachment and incubation in various solutions (*Elodea*), low intensity ultrasound, wounding, light regime, pH, and application of boron and calcium compounds. Callose formation may be reversible if treatment is not too severe. The association of callose with plasmodesmata in parenchyma cells, observed by the above workers, is perhaps suggestive of its ability to block intercellular transport within the parenchyma.

A final factor which could doubtless influence movement of solutes within the leaf parenchyma is the highly variable cellular architecture of this tissue as found in the lamina of different species. In this connection, an unusual layer of paraveinal mesophyll one cell layer in thickness has been reported in *Glycine max* by Fisher (1967). These cells, situated between the palisade and spongy parenchyma and extending horizontally between the veinlets, could conceivably serve a special function in transporting photosynthate or endogenous substances from palisade parenchyma to the veins.

b) *Rôle of the bundle sheath and conductive tissues of leaf veins*

In discussing the transport of organic substances in plants Kursanov (1967) emphasizes the intricate mechanism involved in the transfer of sugars and amino acids from the photosynthesizing parenchyma to the conductive tissues of the fine leaf veinlets. This involves passage through the border parenchyma, or bundle sheath cells. That these cells are indeed somewhat unusual is suggested by several lines of research. The cells would of course act as a final portal in the movement of any pesticide or other molecule before it enters the vascular system and is exported from the leaf. The studies of Fabbricotti-Oberrauch (1965) indicate the high metabolic activity of these cells, and their anamolous chemical nature as compared to other leaf cells. For example, the bundle sheath cells of most species showed an unusually intense storage of crystal ponceau 6R, a sulfo-acid dye of the monoazo group, and of xylene cyanol FF, a sulfo-acid dye of the triphenylmethane group.

Several variations of the mechanism involved in the transfer of carbohydrate assimilates from the leaf mesophyll to the sieve elements have been suggested. It has generally been considered that sugars move in the mesophyll primarily in the form of hexose phosphates. On entering the bundle sheath parenchyma, in which high phosphatase activity has been observed, the phosphate is split off and the hexoses condensed into sucrose, which is subsequently discharged into the sieve elements. Phosphatase activity has also been found in the companion cells of several plant species, and by Braun and Sauter (1964) in groups of cells, the "Phloembeckenzellen", which were inserted as clusters within the sieve tube bundles of Dioscoreaceae. These cells contain nuclei and are rich in plasma and acid phosphatases. The sieve elements also contain some phosphatase activity, especially at the sieve areas. It was considered that these and the "Phloembeckenzellen" take an active part in assimilate transport, being concerned with phosphorylation and dephosphorylation. As has been pointed out by Esau et al. (1957), however, acid phosphatase activity in the phloem and bundle sheath, although suggestive of secretory activity, is not unique to these tissues; it is apparently widely distributed in plants and is a part of the metabolic machinery of virtually all cells.

Brovchenko (1965) has observed that the fine vascular bundles of *Beta vulgaris* leaves contain three to four times more total sugar than the adjacent mesophyll cells. Transport into such bundles must consequently involve considerable metabolic activity. The fact that the fine bundles contained large amounts of monosaccharides, especially glucose, may indicate that the transfer of sugars from the mesophyll to the conductive tissue involves monosaccharides or their phosphate esters. Sucrose synthesis took place within the entire conducting system, but particularly in the finer bundles. Additional evidence suggested that partial or complete hydrolysis of sucrose preceded its entry into the conductive bundles and that, consequently, monosaccharides may compete for entrance. Kursanov et al. (1967) subsequently observed that although continuous hydrolysis and resynthesis of sucrose occurs in the mesophyll, no such "renovation" of sucrose is found within the conductive bundles. Competition for entry into the conducting bundles of *Beta* leaves has been further investigated by TURKINA and SOKOLOVA (1967). They found that sucrose from mixed equimolar solutions of ^{14}C-sucrose and glucose entered the conductive tissues regardless of the glucose. However, once within the tissues, the utilization of sucrose in synthesis of phosphorylated sugars, and of organic and amino acids, was inhibited in the presence of glucose. The glucose content of phloem cells may thus serve as a regulatory factor in affecting the quantity of sucrose available for translocation.

Additional evidence of the involvement of the minor veins in translocation within the *Beta* leaf, and particularly of the importance of the phloem parenchyma therein, comes from three additional reports. The ultrastructural studies of ESAU (1967) demonstrated that these nucleate parenchyma cells are characterized by dense cytoplasm which is particularly rich in ribosomes and mitochondria. The cells were assumed to be well equipped to carry out the enzymatic activities necessary for transformation of sugars moving from the mesophyll into the phloem, and for their secretion into the sieve elements. GEIGER and CATALDO (1969), investigating the translocation of ^{14}C-assimilates in the *Beta vulgaris* leaf by means of histoautoradiography, found that the dense cytoplasm of the parenchyma cells within the minor vein phloem accumulated ^{14}C following application of $^{14}CO_2$ to the leaves. Results suggested that this specialized phloem of the minor veins accumulates sugar and that the parenchyma cells therein cause movement of translocate in the sieve tubes by actively establishing a concentration gradient within the phloem endings. Structural features were compatible with the concept that vein loading precedes translocation. In this connection, the "transfer cells" reported by GUNNING et al. (1968) as a specialized type of phloem parenchyma in minor leaf veins is of interest. These cells, found only in certain plant families, contain a dense protoplasmic content and a large nucleus, and are characterized by numerous protuberances of their inner walls and plasma membranes. The large area of the membrane and the increased ratio of surface to volume suggests a function in accumulating solutes from cell wall fluids, and transferring them to the sieve tubes in the process of exporting assimilates. An alternate role played by these cells may involve retrieval of solutes arriving in the transpiration stream and their subsequent transfer to the sieve tubes.

The phenomenon of vein loading which may occur in conjunction with symplastic movement has been clearly demonstrated by LEONARD and KING (1968) in bean leaves. They placed entire plants in a $^{14}CO_2$ atmosphere in sunlight for ten minutes, following which the leaves were detached for autoradiography. If detached immediately, very little labeling of the veins and petioles occurred, but a strong and uniform labeling of the intercostal areas was found. If, however, the plants were left 24 hours in the dark before the leaves were harvested, the subsequent autograph appeared as a negative of the earlier one: intercostal areas were very light and both veins and petioles were strongly labeled, with the exception of the very young leaves. Vein loading of ^{32}P has been demonstrated in sugar beet leaves by BARRIER and LOOMIS (1957), although loading in this case required a supply of available carbohydrate. Vein loading in conjunction with basipetal transport in detached bean leaves has now been observed by LEONARD and GLENN (1968 a) for ^{14}C-labeled 2,4-D, 2,4,5-T, dicamba, MH, picloram, and assimilates, as well as for ^{32}P. It is of interest that prior treatment with endothall greatly reduced vein loading and subsequent basipetal transport. In contrast to the above compounds which moved into the petiole, diuron moved only acropetally to some extent. Additional evidence (LEONARD and GLENN 1968 b) indicated that the mechanism involved in vein loading is centered principally in the vein margins and endings, perhaps in the bundle sheath cells. Such loading may require that the plasmodesmata in the vein endings and bundle sheath cells remain open. For example, the blockage of such plasmodesmata by endothall-induced callose was found by MAESTRI (1967) to prevent vein loading. The former investigators consider that transport in leaves may be powered by forces in the plasmodesmata of the cell walls between the bundle sheath and the phloem. Since under certain conditions pesticides and other endogenous substances are transported along with assimilates, the above findings are pertinent to their overall foliar absorption.

Although translocation within the stem is not considered in this review, the recent finding of WORLEY (1968) is perhaps pertinent. He observed that movement of sodium fluorescein in fiber cells of pea stem sections was more rapid than in adjacent cortical cells, and was at least partially dependent upon rotational cytoplasmic streaming. The streaming could be reversibly stopped with DNP, and when so stopped, the shape of the advancing dye front was altered and its rate of advance slowed to that of the cortical cells. The behavior suggests the possible importance of these very long fiber cells at least in intracellular transport. That they may serve such a function, wherever present in leaf veins, seems a distinct possibility.

V. Plant factors affecting absorption

a) *Leaf age and development*

As discussed by CRAFTS and FOY (1962), CURRIER and DYBING (1959), and SARGENT (1965), the absorption of most organic and inorganic solutes is generally greater in relatively young leaves. Additional recent research for the most part corroborates this relationship. It is also of interest that the absorption of air pollutants, as indicated by tissue necrosis, is a function of both leaf age and relative age of individual cells within the leaf. In his studies of some 50 species of field plants in the Los Angeles area, NOBLE (1955) noticed that virtually all species exhibited a distinct pattern of banding injury transversely across the leaves following exposure to either natural or artificial smog. The location of the banding on leaves of a given plant was related to leaf age and cellular maturity, maximum injury generally occurring where cellular differentiation had most recently taken place. An interesting corollary is found in the work of PALMER and ENNIS (1960) relating to the penetration of herbicidal oils in hypocotyls of cotton plants. They observed that the intact epidermis of young seedlings, up to an age of about 27 days, was largely impervious to oil. As the plant continued to grow cracks formed in the hypocotyl, primarily as a result of secondary growth of the periderm (cork cambium). Such superficial cracks formed prior to development of cork layers and served as a place for oil to collect and enter. With additional maturity, beyond an age of about 59 days, well developed cork layers formed which again made the hypocotyl impervious to oil. Variations in absorptive characteristics of this nature can have important implications as to the optimum timing of herbicidal applications.

Leaf age apparently affects absorption of inorganic substances in much the same way that it does organic compounds. Thus, AHLGREN and SUDIA (1964) found that absorption of ^{32}P by both leaves and cotyledons of *Glycine max* was greatest in immature leaves and decreased with increasing leaf age until a fairly steady state was reached. Following the absorption of Mg in apple leaves after a spray of five percent magnesium sulfate, OLAND and OPLAND (1956) noted considerable absorption in young leaves but little or none in older ones if sprayed during the day. When treated just before dark, however, old ones absorbed large amounts.

Continued work with growth regulators and related compounds for the most part supports the concept of optimum absorption in younger foliage. Thus, BUKOVAC (1965) found that initial absorption of ^{14}C-3-CP by peach leaves was rapid for the first 12 to 24 hours and then slowly approached a plateau. Young, unfolding leaves showed a greater absorption than did fully expanded or mature leaves. In their studies of absorption and translocation of ^{14}C-GA in *Vitis*, WEAVER *et al.* (1966) observed that leaves about ⅘ full size and still light green afforded some penetration and subsequent translocation, but that little or no penetration occurred in the older leaves. In contrast to the above regulators, foliar absorption of ^{14}C-leucine was found by KAMIMURA and GOODMAN (1964 b) to behave in a somewhat different manner. When this amino acid was applied by means of a glass well affixed to apple leaves, it was found that leaves near the apex absorbed less than those nearer the base. Discontinuities and cracks in the cuticle of the older leaves may have been partially the reason for this phenomenon. Also using apple leaves, WESTWOOD *et al.* (1960) noted that the addition of a surfactant affected the absorption of DNOC more in mature leaves than it did in immature ones. In the case of banana, FREIBERG and PAYNE (1957) found that leaves absorbed urea through the lower surface at the same rate regardless of age, but absorption through the upper surface was more rapid in the older leaves.

In addition to the effect of age upon absorption by the lamina, it may have an additional influence upon transport within the petiole. By measuring the basipetal flux of 2,4-D through petiole segments of primary bean leaves cut from the vicinity of the distal abscission zone and pulvinus, JACOBS et al. (1966) found that transport through pulvinar tissue was less than through petiolar tissue in both young and old leaves. However, flux through the pulvinar tissue of old leaves was much less than in young leaves, whereas flux through petiolar tissue changed little with age. Since no discontinuity of the ^{14}C-label was found in either the petiolar or abscission regions, it was concluded that only the pulvinus limited transport of 2,4-D in the older leaves.

b) *Foliage leaves vs. cotyledons*

Because of anatomical differences and differences in the cuticular structure, it might be expected that absorptive characteristics of cotyledons would be somewhat different from that of true leaves. Such indeed seems to be the case, as reported by a few workers who have examined relative absorption in both of these organs. Relative surface wettability may also vary, as noted by FOGG (1947) in several plant species. He found in general that cotyledons were easier to wet than leaves. BLACKMAN et al. (1958) observed that the triethanolamine salts of several chlorinated phenoxyacetic acids applied in microdroplets on the cotyledons of *Linum* inhibited shoot growth less than a spray of the same quantity applied to the shoot. With sunflower, a drop was more effective placed on the first or second pair of leaves than on the cotyledons. In work reviewed by TSCHIRLEY (1968), the uptake and transport of ^{14}C-labeled urea and 2,4,5-T by the cotyledons and leaves—both primary unifoliate and first difoliate—of *Prosopis glandulosa (P. juliflora* var. *glandulosa)* were followed by autoradiography. The absorption and subsequent transport which occurred within 24 hours was considerably greater in the cotyledons than in either of the foliage leaves, particularly if the latter were not fully expanded. The ^{14}C from applied urea was more mobile than that from the 2,4,5-T, and may have been partially in the form of photosynthates, since urea is apparently hydrolyzed to carbon dioxide and ammonia in the leaves (CLOR 1962), thus making the former compound available for synthesis of sucrose or other photosynthates.

Before the advent of hormone-type plant regulators, ASLANDER (1927) observed that sulfuric acid as a weed spray was injurious to the cotyledons, but not to the pubescent leaves of red clover or the rather waxy leaves of peas. Surface wax structures apparently present only on the leaves were thus able to keep the acid droplets isolated from the actual cuticular surface. The above findings suggest distinct differences between leaves and cotyledons as to retention, absorption, and possibly to relative sensitivity. That diverse sensitivities of these two organs to specific toxicants are due not only to differences in retention and absorption, but also to physiological or biochemical differences, is suggested by HULL and WENT (1952) who observed that when seedlings of alfalfa, sugar beet, endive, spinach, and tomato were exposed to natural smog, or fumigated with an oxidized hydrocarbon aerosol, cotyledons were invariably injured before the foliage leaves. The action on cotyledons was apparently one of oxidation, since a low concentration of ozone alone also attacked cotyledons in preference to leaves. On the other hand, fumigation with sulfur dioxide generally resulted in injury to the leaves before any effect on the cotyledons could be noted. Thus, with the limiting factors involved in retention and absorption being somewhat minimized (as compared to a liquid application), the relative physiological response of cotyledons and leaves seems to be quite different and to be dependent upon the nature of the toxicant.

c) *Upper vs. lower leaf surface and the rôle of stomata*

The generally greater penetration of pesticides and other substances through the lower leaf surface is often accredited to the thinner cuticle and greater number of stomata found on that surface as compared to the upper, at least in most plant species. Several recent investigations, however, have suggested that the relationship may have still additional complicating factors. For example, the greater penetration of certain growth regulators through the lower surface of pear leaves than through the upper, as observed by NORRIS and BUKOVAC (1968), could not be explained solely on the basis of cuticular thickness. This species has a thinner cuticle on the upper surface. Although total quantity of embedded wax is about equivalent in both upper and lower cuticles, the quantity and continuity of molecularly oriented (birefringent) wax is considerably greater in the upper cuticle. This raises the interesting possibility as to whether penetration may be inhibited by the oriented wax to a greater extent than it is by unoriented wax. Additional factors which might favor penetration through cuticle of the lower surface include an increased absorptive area due to corrugation of this surface and possibly a different structure of the epicuticular wax. The studies of DAVIS *et al.* (1968), whereby uptake of 2,4,5-T and picloram was followed by gas chromatography in the leaves of several woody plants, likewise showed greater absorption by the lower than by the upper leaf surfaces. Ratios ranged roughly from 1:1 to 6:1, depending on plant species and herbicide involved.

By means of histoautoradiography PICKERING (1965) studied penetration of 2,4-D, monuron, and dalapon in leaves of bean, cotton, and *Zebrina pendula.* Cuticular entry was observed for all three herbicides in these species, except for *Zebrina,* in which the upper leaf surface was essentially impermeable to monuron and dalapon. With the presence of a surfactant all three herbicides readily penetrated the stomata, which are found only on the lower surface of this species. FREIBERG and PAYNE (1957) observed that stomata were three to four times more numerous on the lower than on the upper surface of the banana leaf, and also found absorption of urea to be more rapid through the former surface. Studying absorption of ^{14}C-leucine by means of its application in a glass well affixed to the surface of apple leaves, KAMIMURA and GOODMAN (1964 b) likewise demonstrated that the lower surface absorbed three to five times as much as the upper.

In the case of the systemic insecticide, Rogor®, DE PIETRI -TONELLI (1965) found that when it was applied as a dilute solution in closely-spaced drops to one surface of a bean leaf, adult mites (*Tetranychus urticae*) feeding on the opposite surface were killed. The fact that mortalities were higher and more rapid following application to the upper leaf surface than to the lower suggested a more rapid absorption through that surface. However, the possibility that mites may be able to penetrate the lower surface more easily and deeply cannot be disregarded. Some rather interesting findings relating the absorption of inorganic ions with leaf surface have also been reported. For example, KANNAN and WITTWER (1967 b), by floating leaves of *Euonymus japonicus* on a solution of ^{59}FeSO$_4$ in the dark at 0° C., found that diffusion of ^{59}Fe through the astomatous upper surface was equal to that of the stomatous lower surface. Although light and a temperature of 20° C. induced opening of the stomata, uptake was enhanced under these conditions by both surfaces. Other work carried out at the same laboratory had shown that chemically induced stomatal closure of excised tobacco leaves in the light had little effect on subsequent rubidium absorption. Using different plant species and a somewhat different technique, EDDINGS and BROWN (1967) also examined the absorption of foliar-applied ^{59}Fe. Absorption proved to be related to stomatal area, and was more than doubled in sorghum but only slightly increased in red kidney bean upon the addition of a surfactant. The greater sensitivity of sorghum in this, and in other experiments involving time-of-day of application, may have been due partially to the alignment of its stomata in longitudinal rows interspersed with veinal tissue. This arrangement, as compared to random distribution in the broadleaf plant, would minimize mean distance the tracer would have to travel from substomatal chamber to the nearest conductive element.

That the degree of stomatal opening is often, but not necessarily always, influential on relative penetration via these structures seems evident. A discussion of stomatal physiology and the regulatory mechanisms which control opening are beyond the scope of this review. These subjects have been considered in detail by KETELLAPPER (1963), ZELITCH (1965, 1967, and 1969), and others. Of particular interest is the recent finding of FISCHER (1968) and FISCHER and HSIAO (1968) that stomata in isolated epidermal strips of *Vicia faba* floated on a potassium chloride solution remain alive and able to respond in a normal manner to light and low concentrations of carbon dioxide. An active accumulation of K^+ was essential to the opening response, and its optimum concentration was dependent upon whether the opening response was mediated by dark or by carbon dioxide-free air; chloride ion seemed only to be of secondary importance in the process.

Many pesticides, because of their volatile nature, penetrate leaves principally by gaseous diffusion. Principles involved in gaseous diffusion through stomata have been discussed in some of the above-mentioned reviews. The interrelationship of diffusion with pore size and spacing was investigated in artificial screens by TING and LOOMIS (1963). They calculated that diffusion through ten-μ stomata spaced at ten diameters was not decreased until the apertures were more than 95 percent closed, and concluded that degree of stomatal opening has no important effect on diffusion from or into a leaf until the stomata are essentially closed. Recognizing the importance of both trichomes and stomata to foliar retention and penetration of pesticides, ORMROD and RENNEY (1968) surveyed these structures on a large number of weeds. Stomatal density ranged from 2.3 to 315.2/mm.^{2} of leaf surface, and lengths ranged from 21.6 to 57.0 μ. Plants grown in partial shade had slightly more stomata than did those grown in full sun. The investigators found a wide range of both stomatal size and density on upper leaf surfaces just within five species of the family Polygonaceae. These species were considered as ideally suited for penetration studies, and should help clarify the questions relating to the role of stomata. That gaseous diffusion through leaf cuticle can indeed occur, even without the presence of stomata, has been demonstrated by HSIEH and HUNGATE (1968). Using isolated cuticles of *Begonia* and *Coleus* leaves, they found that the astomatous upper surface was about as readily penetrated by ^{131}I gas as the lower stomatous surface. A similar behavior has also been reported for carbon dioxide by DOROKHOV (1963). At the content found in normal air, carbon dioxide was absorbed by both upper and lower surfaces of apple leaves, 20 to 30 percent of the total quantity being absorbed by the stomata-free surface. When such leaves, as well as those of oleander and *Ficus microphylla*, were covered on either surface with petroleum jelly, there was a compensatory increase in carbon dioxide absorption by the opposite surface. Studies involving foliar penetration via gaseous diffusion require a knowledge of the relative resistances created by cuticular, stomatal, and boundary-layer effects, and by how these individual components are influenced by environment, species, etc. A detailed discussion relating to the partitioning of total resistance to gaseous diffusion among these various components has been presented by MORESHET *et al.* (1968).

One possible explanation as to why a few experiments show no greater pesticidal absorption through a stomatous leaf surface than through one with no stomata, may lie in the fact that cuticular ledges are present on the guard cells of certain species, which partially or almost completely overarch the stomatal pores. Such ledges have been found by BROWN and JOHNSON (1962) on a number of grasses, but were not present on all of the species examined. These ledges, usually heavily cutinized, have been described by ESAU (1965 a) as occurring on the upper, and sometimes both the upper and lower sides of the guard cells. IDLE (1969) has observed that when scanning electron microscopy is used for the study of stomata having such ledges, surface replication techniques generally fail to represent accurate pore dimensions, as compared to a specimen of the actual leaf itself. It is conceivable that these ledges might not influence gaseous diffusion greatly, but that they might inhibit penetration of pesticide solutions or other liquids seriously, particularly those not formulated with an oil or a surfactant. An additional anomalous cuticular modification which could conceivably influence penetration involves the presence of resinous plugs within the substomatal chambers, such as reported for citrus leaves by TURRELL (1947). Since such plugs apparently did not influence transpiration, it seems reasonable to assume that they would not block stomatal entry via gaseous diffusion; their effect on the entry of liquids may be somewhat more significant.

d) *Moisture stress*

It has generally been conceded that moisture stress inhibits absorption of foliar applied solutes. Research relating to such inhibition has been discussed by CURRIER and DYBING (1959), VAN OVERBEEK (1956), and others. Although there appears to be no question that a favorable water balance is important for optimum translocation, the relationship with foliar absorption is less clear. The cuticle structure, with wax components being embedded in a cutin matrix, would seem to be an ideal mechanism to regulate permeability. Swelling of the polar cutin with an ample moisture supply, as described by VAN OVERBEEK (1956), would theoretically spread the wax components farther apart and thereby enhance the cuticle's permeability, particularly to water and water-soluble solutes. Most recent research with refined experimental techniques, however, does not bear this out, at least in the case of herbicides, with which much of the work has been done. In the work of DAVIS *et al.* (1968 c), also reviewed by TSCHIRLEY (1968), both absorption and translocation of herbicides were followed by a sensitive gas chromatographic assay. Foliar uptake of 2,4,5-T in *Prosopis glandulosa* and *Ulmus alata* seedlings was not affected by four levels of soil moisture stress, ranging from 0.5 to 20.8 atmospheres. Uptake of picloram by *U. alata* was likewise unaffected by stress, although it was significantly inhibited in the stressed *P. glandulosa* seedlings. Translocation of the two herbicides, on the other hand, was considerably reduced even by moderate moisture stress. Using gas chromatographic and thermoelectric techniques in a quantitative study of absorption and movement of these same herbicides in bean plants, MERKLE and DAVIS (1967) observed that the foliar absorption of both compounds was unaffected by extreme moisture stress. Picloram was more mobile than 2,4,5-T at all levels of stress, but final distribution of the herbicides was affected only when stress reached the wilting point. Also using bean, BASLER *et al.* (1961) studied the absorption and translocation of ^{14}C-2,4-D. Like the previous workers, they found that moisture stress had little or no influence on the amount of 2,4-D absorbed by leaves, although it sharply decreased the quantity translocated. PALLAS and WILLIAMS (1962) observed an essentially similar behavior of ^{14}C-2,4-D in bean. The absorption of ^{32}P, however, was influenced by moisture stress; more ^{32}P was absorbed and about eight times as much was translocated below ⅓ atmosphere moisture tension as compared to near three atmospheres. The reason for the similar effect of tension on translocation of the two substances, but not on absorption, is not clear. The fact that the phenoxy herbicide is transported via the symplast, whereas ^{32}P may be transported largely in the apoplast, could partially explain the phenomenon. The complexities imposed by moisture stress have been further demonstrated by the studies of PLAUT and REINHOLD (1967) on movement of ^{14}C-sucrose and tritiated water (THO) in the bean leaf. They found that movement of the labeled sucrose from a small spot on one side of the primary leaf to the remainder of the leaf was far greater in leaves of water-stressed plants than in controls. The *pattern* of movement, however, which was toward the midvein via sideveins, was unaffected by moisture tension. Movement of THO away from a similar spot was just the reverse—far less THO was recovered from the remainder of the stressed leaves than it was from the controls. Girdling experiments suggested that only sucrose movement was via phloem transport, and that such transport was stimulated by water stress. No specific explanation could be found for the completely opposite effects of stress on the two substances. Although TRIP and GORHAM (1968) did not evaluate the effect of moisture stress, they did follow the transport of ^{14}C-glucose and THO within the *Cucurbita melopepo* leaf. By minimizing any differences in initial uptake of these substances by introducing them simultaneously through a cut sidevein or a leaf flap, they observed concurrent transport of both labels down the petiole, with parallel, almost flat gradients. Steam girdling of the petioles blocked the basipetal transport of both labels when the parent compounds were introduced in the described manner. Such blockage, along with the observed concurrent transport of ^{14}C-sugars, THO, and T-photosynthate in constant ratio, suggests that the solution flow of sugar cannot be excluded as a transport mechanism, assuming that THO moves as liquid water in the phloem. Although these reports at first appear contradictory as far as transport of THO vs. sugar is concerned, they perhaps emphasize the differential effect imposed by moisture stress on transport of the two compounds—stress was adjusted only in the former experiment. Also, a different sugar and a different plant species was involved in each case.

e) Physical nature of the surface and extraneous matter thereon

The affinity of a leaf surface for particulate matter from sprays or other atmospheric fallout would seemingly be influenced by degree of surface roughness. This influence, with both insect and leaf cuticles, has been discussed in detail by HOSKINS (1962). It has also been substantiated by the work of GÜNTHER and WORTMANN (1966). Their ultrastructural studies of leaf surfaces demonstrate that particles adhere with little tenacity to smooth, waxy-surfaced cuticles, such as found on *Crambe maritima*, whereas rough-surfaced leaves such as *Taraxacum officinale* may exhibit a strong affinity for aerosol particles. Variable foliar affinity of this nature could probably affect adherence not only following pesticide sprays, but also following fallout of industrial or radioactive dusts. Such residues could conceivably be minimized by genetic selection of food and forage crops, and perhaps by pretreatment with specific protective compounds of one type or another. In examining the foliar uptake of diquat and paraquat by several plant species, BRIAN (1967 b) noted that these herbicides also became adsorbed to extraneous matter on the leaf surface. Thus, soiled or dusty leaves would very likely impair the accuracy of quantitative or autoradiographic studies.

f) Species and varietal differences

Various degrees of susceptibility to specific chemicals are often found among different plant species and varieties. Because of the design of many experiments, it is often difficult to establish whether these differences are due to variable penetrability of the chemical involved or to differences in its metabolism within the cell. Experiments with radioactive isotope-labeled compounds have helped clarify some of the questions. For example, varieties of *Lotus corniculatus* vary in their susceptibility to 2,4-D. BLACKLOW and LINSCOTT (1968) found that when the potassium salt of 2,4-D-2-^{14}C was applied to the second and third oldest leaves of this plant, absorption, as determined by retention on the seventh day, was not significantly different between the susceptible variety "Viking" and a resistant intercross. The variable susceptibilities were apparently not due to absorptive differences. Absorption in this species was virtually complete within 24 hours. That absorptive differences between species exist in the case of 2,4-D is also known, as exemplified in the studies of ZEMSKAYA and RAKITIN (1967) on the final localization of this compound in maize and sunflower leaves. Differential centrifugation of the maize leaf homogenate indicated that 2,4-D became localized in the plastids, nuclei, mitochondria, proteins of the supernatant, and particularly within the cell walls. In contrast, 2,4-D localization in the sunflower leaf preparations, even after several different modes of introduction, was detected only in the supernatant. The authors concluded that 2,4-D was bound by proteins of the protoplasmic membranes and cellular organelles of the maize leaf cells, but existed in the sunflower leaf cells essentially in a free state.

It is known that *Pinus ponderosa* and *Abies concolor* differ considerably in their susceptibility to amitrole. Whether this is due principally to variable degrees of absorption in the two species or to detoxification via different metabolic patterns has been examined by LUND-HØIE and BAYER (1968). Although amitrole was found to be translocated and metabolized similarly by both species, it was determined that the greater susceptibility of the pine was due to a lower resistance to uptake into the needles. In studies relating to the differential phytotoxicity to ioxynil to mustard, barley and pea, DAVIES *et al.* (1968) initially demonstrated that relative foliar retention in these species was a major factor in determining sensitivity, although it was not the sole factor involved. They subsequently found that penetration of ^{14}C-ioxynil into leaf disks or segments was considerably greater with the sensitive mustard than with either barley or pea, and was unrelated to stomatal density. When plants were sprayed, only by increasing the concentration of ioxynil and adding a surfactant could almost as much be made to enter the resistant barley as entered the mustard at a lower concentration without a surfactant. That some of the observed varietal differences in susceptibility to certain toxicants are related to absorption has also been demonstrated. Thus MILLER and ABOUL-ELA (1968), using microscopic time-lapse photography and autoradiography, found that the cotton defoliant $(BuS)_3PO$ readily entered the upper leaf cuticle. Entry was accompanied by partial plasmolysis, cuticular vesication, vesicle collapse, and progressive blackening of the palisade cells. A quantitative visual determination of the number of cells injured by this rapid process indicated that differences between the susceptible Deltapine 15 variety and the resistant selection G-272 were due to absorptive differences.

VI. Environmental factors affecting absorption

The influence of various environmental factors on absorption and
transport of the dipyridylium compounds, diquat and paraquat, has
recently been reviewed by Akhavein and Linscott (1968) in this book
series, and by Calderbank (1968). It will not be further considered
here, except for the effect of light. Information on other pesticides is
somewhat more scattered and will be discussed individually under the
specific environmental factors involved.

a) *Temperature*

Since it is generally conceded that absorption of most foliar-applied
compounds is governed by both metabolic and non-metabolic components, it seems likely that the overall process would be accompanied
by fairly high temperature coefficients. That this is indeed the case is
suggested by the majority of research in which the effect of temperature on absorption is considered. Physiological explanations as to the
most likely causes for such Q_{10}'s is capably discussed by Sargent
(1965) and several of the other previously-mentioned reviews. Additional recent research generally corroborates the above relationship, a
few examples of which will be cited.

In studies on the movement and metabolism of ^{14}C-2,4,5-T in
Prosopis juliflora seedlings, Morton (1966) found that no differences
in foliar absorption occurred at 70° as compared to 85° F.; however, a
significant increase was found at 100° F. Approximately half of the
isotope applied to a single leaf was absorbed over a 72-hour period.
In following the absorption and translocation of ^{14}C-atrazine in *Agropyron repens* by autoradiography, Wax and Behrens (1965) observed
a slightly but not significantly greater uptake and acropetal transport
of the foliar-applied compound at 80° than at 60° F. Prasad *et al.*
(1967) did find about a four-fold greater foliar absorption of ^{14}C-
dalapon in bean when post-treatment temperature was maintained at
43° as compared to 26° C, with relative humidity about 60 percent in
both cases. The combined effects of temperature and light intensity on
MSMA absorption and transport in *Sorghum halepense* have been
studied by Kempen and Bayer (1969). They noted that the time required to give 50 percent necrosis of rhizomatous foliage following
application to the leaves was nearly 12 days at 15° C. and 310 footcandles, whereas the same degree of necrosis was achieved in less than
one day at 35° C. and 2,800 footcandles. Even root temperature may
influence foliar penetration, at least in the case of ^{32}P, as demonstrated

by PHILLIPS and BUKOVAC (1967). It was established that this isotope, applied to bean and pea leaves, was absorbed and transported to the roots to a greater extent when the latter had been maintained at 18° to 24° C. than when maintained at 7° to 13° C. That the effect was principally on absorption, rather than translocation, was suggested by the fact that leaves excised from plants grown one to four days with roots maintained at the above temperatures continued to absorb ^{32}P at rates which were a function of the temperatures to which the roots had been exposed. ^{45}Ca was apparently absorbed by the leaves to some extent in these studies, but its export from the treated leaves was negligible.

b) *Humidity and rain*

Although a few experiments suggest that exceptions exist, foliar absorption of both organic and inorganic substances is generally facilitated by a high humidity. Not only does this condition favor stomatal opening, thereby enhancing penetration via this route, but it also slows the drying of spray deposits, thus extending the time available for absorption. Although humidity influences degree of hydration of the leaf cuticle, which in turn may influence its permeability, such hydration is perhaps more a function of available soil moisture than of atmospheric humidity under most conditions. This has already been discussed under moisture stress.

The studies of PRASAD *et al.* (1967) with both ^{14}C- and ^{36}Cl-labeled dalapon on a number of different plant species demonstrated that foliar absorption and translocation is greater at high (88 percent) posttreatment relative humidity (r.h.) than at medium (60 percent) or low (28 percent). Although the humidity effect was considered to be largely due to drying rate, when plants held at low humidity following treatment were periodically rewetted at the area of application, absorption was enhanced somewhat but never to the extent achieved with the high humidity treatments. If plants were actually grown under either a 95 or a 28 percent r.h., but both treated under the high humidity, absorption was markedly greater in plants grown under the high humidity. It is of interest that either a high ambient humidity or rewetting of the application area under lower humidity conditions promoted uptake through both stomatous and astomatous leaf surfaces. Investigating the movement and metabolism of ^{14}C-Alar (B-9) in several plant species, SACHS *et al.* (1967) noted that 65 percent of the quantity applied penetrated within 24 hours at a 95 to 100 percent r.h., whereas only about two percent penetrated at a 50 to 60 percent humidity. The penetration of this growth retardant was most rapid during the first 12 to 18 hours, but continued at a much reduced rate for at least seven days. Freiberg and Payne (1957) likewise found that 65 percent of the urea applied to banana leaves was absorbed within 25 minutes under humid conditions, but that under drier conditions it took 24 hours for 62 percent to be absorbed. In contrast to the above absorptive behavior, Morton (1966) could detect no significant differences in foliar absorption of ^{14}C-2,4,5-T in *Prosopis juliflora* seedlings at four r.h. levels ranging from 35 to 100 percent. This anomalous response was considered possibly related to the fact that *Prosopis*, a relatively xerophytic species, may be better able to adapt to low humidities than certain other plants.

The effects of environment upon foliar absorption of DNOC have been studied by Westwood *et al.* (1960). This compound, which has had various uses as a spray for the chemical thinning of fruit, as an ovicide in the treatment of dormant fruit trees, and as a contact herbicide, was absorbed by apple leaves to an extent which was directly related to duration of high humidity pretreatment. Such humidity-enhanced absorption, as measured by subsequent leaf necrosis, was minimized by artificial rain following the spray application. Whether rain will enhance the penetration of a pesticide or wash most of it off depends upon the quantity of rain involved, the time interval at which it occurs following the spray application, the solubility characteristic of the pesticide, and the physical nature of the leaf surface. Thus it is conceivable that the natural wetting of a leaf by dew or rain could either increase or decrease the quantity of a pesticide which is finally absorbed. Linscott and Hagin (1968) found that simulated rainfall of 0.1, 0.5, and 2.0 inches applied to various forage crops at intervals up to seven days following treatment with the dimethylamine salt of 2,4-DB resulted in an average herbicidal loss from the foliage of 21, 60, and 93 percent, respectively.

When a light dew or a rainfall insufficiently heavy to cause foliar runoff occurs, enhanced absorption of spray deposits may occur. For example, Bukovac (1965), examining the absorption of 3-CP by peach leaves, observed that when the residue on the leaf surface was simply rewetted with deionized water every two hours over a 12-hour period, overall absorption was increased by 47 percent. Absorption of inorganic substances may follow a similar pattern. Thus, Ambler (1964) found that the maximum amount of ^{85}Sr translocated out of bean and maize leaves amounted to about 0.6 and 0.5 percent, respectively, of the quantity applied, if the leaves were periodically rewetted. If they were allowed to air dry and stay dry, transport out of the leaves was generally less than 0.01 percent of that applied. These differences were presumably largely due to differences in absorption. In contrast to such a response Oland and Opland (1956) observed that the uptake of magnesium by young apple leaves, following a spray of five percent magnesium sulfate, was not increased by rewetting.

With significant quantities of rain, as described above, active leaching can of course occur. TUKEY *et al.* (1965) describe leaching of inorganic minerals, carbohydrates, amino acids, and organic acids from the foliage of numerous diverse species by the action of rain, dew, and mist. MORELAND *et al.* (1966) observed that organic substances are released to the external environment by foliar leachates, as well as by direct root secretions, and that such substances could function endogenously as inhibitors and regulators of seed germination, dormancy, root initiation, and various other processes. Endogenous gibberellins, as well as GA absorbed by five-cm. stem segments and subsequently translocated to attached leaves, were shown by KOZEL and TUKEY (1968) to be leached from the foliage of *Chrysanthemum morifolium.* Since pesticides, as well as endogenous substances, may be leached from one plant and subsequently absorbed by the foliage or roots of others, the phenomenon is of particular importance to the present discussion.

Leaching tends to increase with increased maturity of the leaves, reaching a peak at senescence. In studies with ^{45}Ca and ^{86}Rb, TUKEY *et al.* (1965) noted that leaching was relatively independent of light, air temperature, and volume of the leaching solution. It was also observed (TUKEY 1966) that leaves with a smooth waxy surface (e.g., pea, sweet orange, sugar beet) which are wetted with some difficulty, are less subject to the leaching action. Flat-surfaced, pubescent leaves (e.g., bean, squash) which are readily wetted, leached with facility. No correlation of mineral nutrient loss by leaching with stomatal density could be detected; in fact leaching could be as great or greater from astomatous leaves as from those with stomata. The leaching of inorganic cations is apparently strongly related to nutrition. For example, HOHLT and MAYNARD (1967) found significant amounts of magnesium and calcium leached from leaves of intact spinach plants grown at four meq. Mg^{++}/l., but not when grown at 0.5 meq./l. With isolated cuticles, influx and efflux rates for Mg^{++} were related to stomatal frequency. This is in contrast to the above-described finding with intact plants, but does suggest that stomata may exert some influence on leaching. Another mechanism whereby materials may be exported to the leaf surface and leached therefrom is via guttation fluid. ZIEGLER and VOGT (1968) found that when ^{14}C-flurenol was applied to the leaves of young oat seedlings, the guttation fluid subsequently given off contained considerable quantities of labeled material consisting of the unchanged compound and derivatives of various R*f* values. Ionic exchange may also be involved in the process of leaching. For example, MECKLENBURG *et al.* (1966) demonstrated that the distribution of ^{45}Ca within the bean leaf and its movement through the cuticle are dependent upon exchange phenomena, involving exchange sites on the pectinaceous materials permeating the cuticle and on the cell walls. Subsequent leaching then follows either an exchange by hydrogen from the leaching solution and/or a diffusion of ions from the translocation stream of the foliage.

c) *Light*

A decade ago CURRIER and DYBING (1959) remarked that evidence for the effects of light on permeability of cuticle and protoplasm could not currently be assessed. The statement is still true. Reports relating to light effects on the absorption of pesticides and various other organic and inorganic substances continue to be conflicting. For one thing, it is difficult to adjust light intensity or quality over very extreme ranges without affecting, at least to some extent, the temperature and humidity. The latter factors do of course influence absorption. It may also be influenced, indirectly, by certain other light-related factors. For example, spray deposits remaining on the leaf surface for any great length of time may suffer loss due to photodecomposition, largely from ultraviolet light. Pesticides, however, do vary considerably in their susceptibility to photodecomposition. Also, the products of such degradation could still be absorbed. In considering overall penetration into the leaf, the component portion which occurs via the stomata may be influenced by light, in that light generally effects an opening of these structures. Finally, in enhancing the production of photosynthate within the leaf, light may influence the movement of exogenous materials, since it is known that the transport of such materials is related to the movement of sugars and other assimilates. Such an effect would probably be more on intercellular movement within the mesophyll and fine vascular elements of the leaf than on initial penetration of the cuticle.

The complex role played by light is well exemplified in the comprehensive studies of SARGENT and BLACKMAN (1965) on ^{14}C-2,4-D penetration into bean leaf disks. Within certain concentration limits, and between zero and 1,000 footcandles of light, they found a small increase in penetration rate into the lower surface which was proportional to concentration and continued over 56 hours. At intensities in excess of 1,000 footcandles the slow, steady increase also occurred, but was followed after a few hours by a second phase in which the penetration rate greatly accelerated. The latter phase did not occur at 1° C., and was reversed if disks were transferred to darkness. Likewise, it did not occur under any conditions in the upper leaf surface. When leaf disks of *Ligustrum ovalifolium* were used, penetration into both the astomatous upper and stomatous lower surfaces remained constant, but was progressively more rapid from zero to 2,000 footcandles. No delayed "surge" of penetration occurred through eighter surface. Previous work, involving the temperature coefficient relationships of 2,4-D penetration, suggested the relative importance of physical factors in determining penetration into the *Ligustrum* leaf. Q_{10} studies with the *Phaseolus* leaf, in combination with the second accelerated phase of 2,4-D entry, may indicate a relatively greater involvement of a metabolic component which could become limiting under certain conditions. The studies of AHLGREN and SUDIA (1967) on absorption of phosphate by *Glycine max* leaflets also show a light-induced enhancement. Maximum absorption occurred with immature leaflets of plants treated in the light; minimum absorption occurred with the same leaflets treated in the dark. When leaflets were treated in the light but covered with opaque material, relative age of the leaflets exerted a profound effect on absorption. Thus, mature leaflets absorbed only as much as those of dark-treated plants, but immature leaflets so treated absorbed almost as much as immature leaflets treated in the light. If leaves below the covered treated leaflets were excised, the enhanced phosphate uptake did not occur, thus suggesting that the lower leaves transported assimilates or other substances responsible for the enhanced uptake to immature but not to mature leaflets. It is of interest that 2,4-DNP inhibited uptake in both light and dark, although light partially overcame the inhibition.

In contrast to the above-described positive effects of light on 2,4-D and phosphate uptake, R. C. BRIAN (1967 b) found that penetration of diquat and paraquat into leaves of intact *Beta vulgaris* and tomato plants in the dark exceed that in the light. Such behavior suggested that entry was not through open stomata, but through the cuticle. Darkness for as little as four hours after treatment increased uptake by tomato almost twofold. DAVIS *et al.* (1968 a) studied foliar uptake of 2,4,5-T and picloram in *Quercus virginiana* and other woody species by gas chromatographic analysis following application of one-μl. droplets to the surface of detached leaves. They noted that absorption of these compounds was not significantly affected by light wave lengths from three different portions of the visible spectrum; however, absorption of picloram was inversely related to light intensity, in agreement with the above-described behavior of diquat and paraquat. In studies involving the absorption, transport, and fate of ^{14}C-dicamba in *Cyperus rotundus*, MAGALHAES *et al.* (1968) observed that exposure of the plants to a light intensity about four percent of full sunlight for 30 days prior to treatment enhanced translocation, apparently due to favored penetration through the thinner cuticles which such plants developed. It should be kept in mind that both the phytotoxicity of the active constituent and the concentration at which it was used vary widely in the above experiments. Processes governing the absorption of highly toxic materials, accompanied by visible foliar injury, are no doubt very different from those governing absorption of nontoxic organic or inorganic compounds. Such differences may partially explain the diverse effects of light on absorption which are found in the literature.

VII. Properties of the pesticide formulation as related to retention and entry

a) *The active ingredient*

1. Molecular structure, chemical and physical properties.—Because of the highly polymerized nature of the outer portion of the cuticle, and its high wax content in most plants, it would be expected that relatively nonpolar compounds having semi-lipophilic characteristics could penetrate this barrier most effectively. Research with all types of pesticides indicates, with few exceptions, that this is indeed true. Penetration, followed by hydrolysis, often results in an active moiety having greatly increased water solubility. Thus penetration in varying degrees through the lipophilic portion of the cuticle may be achieved with the intact molecule, followed by hydrolysis and further transport via aqueous routes. When molecules are relatively resistant to hydrolysis or other metabolic breakdown, it is particularly important that they be of the optimum polar-apolar balance in order to accomplish their desired effect. Thus, with systemic insecticides or translocated herbicides, effective accumulation throughout the plant is dependent on both adequate absorption and translocation. Should the polar-apolar balance be sufficiently extreme to inhibit either one of these processes excessively, the overall desired activity may not be achieved. The relationships of these and other molecular characteristics, as they relate to absorption of various organic substances, have been discussed by HARTLEY (1966), HOSKINS (1962), and in several of the previously-mentioned reviews. LINSER (1964) also emphasizes the importance of these characteristics in his treatise on the design of effective herbicides.

In contrast to the above generality that a high lipophilic tendency facilitates foliar absorption, there seems to be some indication that such absorption is also positively related to water solubility of the penetrant. For example, SMITH *et al.* (1967) found that uptake of ^{36}Cl- and ^{14}C-labeled Dursban® following topical or spray application to leaves of several plant species was insignificant, amounting to one to two percent. This insecticide has a water solubility of only about two p.p.m. On the other hand, the plant growth retardant Alar, which is soluble to the extent of about ten percent, shows particularly good penetrability. EDGERTON and GREENHALGH (1967) also noted that when this compound (^{14}C-labeled) was applied as a spray to the foliage and young fruit of apple trees it was readily absorbed and translocated, as determined by assay of various parts of the fruit and of the branches during the subsequent dormant season. Of course different species are involved in the above experiments, and quantitative comparisons cannot readily be made, but the highly soluble Alar apparently was very readily absorbed and transported. The same relationship of penetrability with water solubility is evident in the studies of Foy (1964 b) with cotton. He found that ^{14}C-prometone and five other alkylamino-s-triazines pentrated the foliage in direct relationship to their water solubilities, which ranged from five to 750 p.p.m. Accumulation of ^{14}C occurred in a wedge-shaped pattern distal to the point of application on the lamina, suggesting apoplastic movement with the transpiration stream.

The fact that foliar absorption and transport are functions of several factors other than water solubility is also quite apparent. WILLIAMS et al. (1967) applied either ^{14}C-labeled sorbitol or sucrose to the lower surfaces of two central leaves of cotton seedlings. Although both compounds are highly soluble in water, the polyhydric sugar alcohol was absorbed and transported out of the treated leaves within 24 to 96 hours at about twice the rate of sucrose. PLUMMER and KETHLEY (1964), using isotope-labeled amino acids from insect serum, hydrolyzed proteins, or mixed solutions of purified biochemicals, examined their relative absorption by the sarcophageal (insect-digesting) region of leaves of the pitcher plant (Sarracenia flava). Although the amino acids used have a variable but quite moderate degree of water solubility, they varied substantially in rate of accumulation within the leaves. The complex interplay of molecular volume, oil/water partition coefficient, polarity, and other factors which affect molecular penetrability, and which perhaps best explain some of the apparently anomalous results noted above, have been comprehensively described by COLLANDER (1959). In this connection, the evaluation by MILBORROW and WILLIAMS (1968) of penetration of certain toxic non-electrolytes into Nitella cells is of interest. The relationship of penetration with partition coefficient has apparently proven to be somewhat variable, and to be dependent upon the technique used. Thus, when penetration is determined as a function of subsequent biological activity, penetration is generally found to increase with increases in partition coefficient up to a maximum, after which it decreases. On the other hand, direct measurements of penetration have indicated a direct proportionality to partition coefficients. However, if penetration rates measured with the latter technique are described by an equation which includes a quadratic function of the partition coefficient (within limits), there appears to be no incompatibility between penetration determined by biological activity as compared to that obtained by direct measurement. As to investigations such as described above, it should be recalled that penetration is measured principally in the cellular membrane; they are thus applicable only to that particular component of overall foliar absorption.

In the case of the phenoxyalkyl acid herbicides, structural changes in either the ring or side chain of the molecule can markedly alter absorptive properties. For example, KIRKWOOD et al. (1966) and ROBERTSON and KIRKWOOD (1966) demonstrated with ^{14}C-labeled compounds that the high susceptibility of Vicia faba to MCPA was related to the rapid cuticular penetration of this molecule and its extensive movement throughout the plant, whereas the resistance to MCPB was associated with its virtual confinement to the treated leaves. Such selective adsorption of the latter compound apparently took place at protein or vacuolar sites near the point of entry, thus minimizing quantities which might reach sites of β-oxidation. In distinct contrast to the behavior in higher plants, MCPB was more toxic than MCPA to bacteria, fungi, and algae. It was also more rapidly absorbed by mycelial pads of *Aspergillus niger* than was MCPA, thus suggesting that differential absorption is a factor influencing the relative toxicity of these two compounds. Studying foliar absorption and translocation of several formulations of ^{14}C-labeled 2,4-D and 2,4,5-T in *Acer macrophyllum* seedlings, Norris and Freed (1966) found that absorption increased with decreasing polarity of the molecule. Thus, the relatively lower mobility of the esters within the plant was offset by their greater absorption, thereby resulting in greater accumulation in the roots as compared to other formulations. By means of both liquid scintillation counting and gas chromatographic techniques, Morton *et al.* (1968) likewise observed that leaves of *Prosopis juliflora* absorbed approximately two to three times more 2,4,5-T when applied as a butoxyethyl ester formulation as compared with the ammonium salt. The apparently limited translocation of the ester formulation within the plant, however, resulted in roughly equivalent accumulations of the two formulations in the stem.

That substitutions within the ring may also alter absorptive properties has been shown by several investigators. Additional halogens in a molecule generally increase its lipophilic characteristics, which should theoretically enhance cuticular absorption. As an indicator of relative adsorption of phenoxyacetic acids within plant tissue or on soil particles Leopold *et al.* (1960) investigated adsorption of some 17 of these compounds onto carbon. They found adsorption to be generally proportional to the number of chlorine substitutions in the phenyl ring and observed a strong inverse correlation between adsorption and water solubility. As previously discussed, there is good evidence that herbicidal selectivity is at least partially related to these differences in foliar adsorption, which can occur among species as well as among molecules. Exceptions to the positive relationship between chlorine substitution and apparent foliar absorption do exist and have been discussed (Hull 1964 a).

2. **Concentration interactions.**—Some of the intricacies involved in the penetration of certain organic and inorganic substances into individual cells have already been described. Of the numerous factors affecting such penetration, concentration is of utmost importance. Thus, of the specific absorptive mechanisms involved, each is generally operative only within a specific concentration range, which in turn is dependent upon toxicity and other factors. Such cellular penetration is of

course only one facet of overall foliar absorption. It is the contention of many investigators that herbicides which have a high contact toxicity, or are used at excessively high concentrations, cause injury to the conductive tissues within the leaf and thereby inhibit further transport of the toxicant out of the leaf. For example, VOEVODIN and ANDREEV (1960), studying the absorption and translocation of ^{14}C-labeled 2,4-D (sodium salt) in several weed species, found that high rates (e.g., 1.5 kg./ha) often impaired absorption and translocation, thus preventing destruction of the roots. BADIEI et al. (1966) examined the absorption and transport of ^{14}C-2,4,5-T (butoxyethanol ester) applied to individual leaves of Quercus marilandica seedlings. As the quantity in a single ten-μl. droplet was increased from ten to 25 to 50 μg., absorption (as a percentage of that applied) decreased significantly with each increment. Translocation out of the leaf also decreased, but only as the quantity was raised from 25 to 50 μg. In the chemical control of Typha angustifolia with 2,4-D, LEVI (1960) recognized that above a certain threshold concentration, excessive contact kill resulted in an insufficient quantity arriving in the tissue at some distance from the point of application.

Foliar absorption is not only a function of the toxicant concentration, but also of the interaction of such concentration with that of specific constituents of the carrier, notably surfactants and oils. In a study involving the action of the triethylamine salt of 2,4,5-T on Prosopis juliflora seedlings (HULL 1957), microscopic injury to the phloem was evident at some distance from the point of treatment when the chemical was applied at a low concentration in combination with a relatively low surfactant concentration. With high concentrations of either 2,4,5-T or surfactant, and particularly with high concentrations of both, marked contact injury occurred but such proliferation of the phloem and accompanying crushing of the sieve tubes and companion cells at loci distant from the point of application was minimized or did not occur. A somewhat related behavior was observed by WEBSTER (1962) with detached Kalmia angustifolia leaves. While evaluating entry of ^{14}C-2,4-D at various ratios of 2,4-D to surfactant, he found that the entry rate into mature leaves was positively correlated with surfactant concentration but that entry into young leaves was negatively correlated, within the range of ratios used.

b) The influence of adjuvants

1. Surfactants.—Of the numerous types of adjuvants which may be included in a pesticide spray, a surface-active agent is perhaps most consistently effective in enhancing foliar absorption of the active toxicant and thereby promoting its maximum effect. As a single example, CORNS and DAI (1967) found that the saplings of Populus tremuloides and P. balsamifera could be killed as readily with sprays containing

500 p.p.m. of picloram plus one percent of a nonionic blended surfactant as they could with 2,000 p.p.m. of picloram without the surfactant. Esters of 2,4-D and 2,4,5-T were less effective than picloram, but at 1,000 p.p.m. with the surfactant they were as effective as 2,000 p.p.m. without it. A hundred additional recent examples of such surfactant-induced enhancement could probably be cited, some even more spectacular, and including all classifications of pesticides. Against these positive effects, which can be demonstrated by numerous techniques, there are a considerably smaller number of reports indicating that surfactants are without effect or may even inhibit penetration of the active ingredient. In the remainder of this section we shall try to examine some of the factors which influence the manner in which a surfactant functions, and which may determine whether or not it can be expected to enhance absorption of the pesticide involved. If the desired effect of a specific insecticide, fungicide, or herbicide can be achieved at a substantially reduced rate when used in combination with a surfactant, the possibilities of environmental contamination are of course proportionately minimized. As a group, surfactants have a far lower mammalian toxicity than do the majority of pesticides.

Surfactants may be classified as nonionic, anionic, or cationic. Amphoteric blends of various ionic-nonionic combinations are of course possible. Some of the physical and chemical properties of these classes have been discussed by Behrens (1964). He and also Foy and Smith (1969) consider the nature and mode of action of surfactants which are useful with herbicides. Surfactants particularly effective in combination with the dipyridylium compounds, diquat, and paraquat, are described by Akhavein and Linscott (1968). The term surfactant is generic in nature, and includes surface-active substances which are often specifically designed to serve as a detergent, an emulsifier, a wetting agent, or a spreader-sticker. Although these properties are mutually desirable, any one or all of them may not necessarily be correlated with what might be called biological enhancement, i.e., the promotion of maximum absorption of the toxicant. For example, a compound capable of maintaining an excellent oil/water emulsion may not effect absorption of the associated toxicant significantly. Smith and Foy (1967) examined the ability of ten surfactants of differing ionogenic classes to enhance the activity of paraquat on maize. Although solutions of the individual surfactants differed considerably in their surface tension and pH values, neither of these properties showed any correlation with the paraquat-enhancing capabilities of the surfactant. In spite of these diverse physical properties, Foy and Smith (1965) further showed that all surfactants examined (seven) markedly enhanced the activity of dalapon on maize, in proportion to their concentration. Minimum surface tensions and contact angles occurred at 0.1 to 0.5 percent concentration, but maximum herbicidal activity occurred at ten times these levels or greater. Thus, above the low concentrations, herbicidal enhancement was not correlated with surface tension lowering, contact angle, wettability, or initial toxicity.

Although most work has shown that penetration enhancement of the toxicant is proportional to surfactant concentration over a relatively great range, this relationship does not hold for all circumstances. EVANS and ECKERT (1965), investigating various paraquat-surfactant combinations on *Bromus tectorum*, obtained maximum penetration and activity at surfactant concentrations of only 0.06 and 0.12 percent; higher concentrations decreased the activity. One of the reasons why surfactant concentrations exceeding a certain threshold level may cause trouble (PARR and NORMAN 1965), is because of the formation of molecular aggregates or micelles. This critical micelle concentration (cmc) is associated with the maximum potential for lowering of surface and interfacial tensions. The aggregates usually have hydrophobic groups oriented toward their periphery, and hydrophilic groups toward the inside. Their molecular weights may be as high as 100,000. Many surfactant-induced biological effects are not related to surface tension changes below the cmc range, but are related to conductivity, osmotic, and electrophoretic characteristics, and certain other physical-chemical properties which are influenced by concentrations in excess of the cmc.

As spray adjuvants, surfactants are sometimes used at concentrations as high as one percent or more. Such concentrations may be desirable where the objective is to enhance pesticidal penetration; however, in biological experimentation involving complex growth phenomena as influenced by certain regulators, etc., such concentrations may adversely affect the outcome of the experiment. These effects, which are due to the phytotoxic characteristics some surfactants possess even at very low concentrations, have been emphasized by PARR and NORMAN (1965). They also review some of the biological effects which may occur from surfactants *per se*, when applied to plants at the low concentrations of 0.0005 to 0.1 percent. VIEITEZ *et al.* (1965) found that Tween 80 enhanced the elongation of *Avena* coleoptile sections slightly at concentrations from 0.001 to 0.1 μg./ml., whereas Tween 40 had little effect on elongation at any concentration. Tween 20, on the other hand, strongly inhibited elongation at all concentrations. All three of these nonionic surfactants have basically similar solubility characteristics, and are relatively hydrophilic. The surfactants vary only in the fatty acid moiety of their molecules, which apparently accounts for the growth enhancement or inhibition. The responses were also evident in the presence of IAA, there being some indication of antagonism between this regulator and Tween 20.

Although the growth repressive effect described above is probably largely due to inhibition of cell elongation, there is evidence that surfactants also inhibit cell division. Thus, NETHERY (1967 a and b) examined the effect of 22 surfactants, including nonionic, anionic, cationic, and amphoteric classes, on mitotic inhibition of the primary pea root meristem. Significant inhibition occurred at a concentration of 0.1 percent with 16 of the compounds, with no relationship between inhibition and ionogenic class being evident. Of six known biodegradable surfactants, he found five to be toxic at one percent. It is of interest that SMITH et al. (1967), working with four new biodegradable surfactants, noted that they were inherently phytotoxic but gave excellent visual wetting of maize foliage. As to enhancing the herbicidal activity of amitrole, dalapon, and paraquat, they proved to be equal or superior to some of the standard surfactants in common use. Of some 20 surfactants of different ionogenic classes, the ionic types were found by STROEV (1965) to sharply increase the intracellular catalase activity of a suspension of Candida albicans cells. Agents capable of such activation were also found to arrest growth of the cultures, these effects apparently being related to an increased cell wall permeability. Nonionic surfactants did not produce the effect. One possible additional mechanism whereby surfactants exert their toxic action is suggested by VERNON and SHAW (1965): at only 0.007 percent in a suspension of spinach chloroplasts, Triton X-100 caused uncoupling of photophosphorylation and resultant stimulation of electron transfer reactions. Rates of this biodegradable surfactant in excess of 0.02 percent induced certain additional anomalies in photochemical processes. In general, it is desirable to know the inherent phytotoxicity of a surfactant, since one may be hoping to prevent foliar damage in the use of a systemic insecticide in one case, or attempting to create maximum plant damage with a herbicide in another.

As far as biological enhancement of pesticides is concerned, one ionogenic class is generally not greatly superior over another, although exceptions to this statement have been demonstrated; certainly there is great variation within individual classes. In experiments involving the foliar nutrition of coffee seedlings BOROUGHS and LABARCA (1962) found that uptake of ^{32}P-labeled potassium or ammonium dihydrogen phosphate occurred in the presence of any ionogenic class of surfactant. Measurements two days following treatment of a single leaf indicated both anionic and cationic types to be superior to nonionic in enhancing ^{32}P uptake. However, by the 18th day there were no significant differences in degree of enhancement among the three surfactant types. BAYER and DREVER (1965) evaluated the toxicity of diuron to oat plants at various intervals up to 20 days following treatment. They likewise found that diuron absorption, as reflected by the toxicity index, was essentially independent of surfactant type—nonionic, anionic, or cationic. They noted considerable variation among individual surfactants, and obtained consistent phytotoxic effects only at surfactant concentrations above 0.1 percent. Toxicity was, in fact, more a function of surfactant concentration than of diuron concentration.

The relationship of the solubility characteristics of the toxicant to its penetration has previously been mentioned. One mechanism whereby surfactants exert their influence is by enhancing such solubility. Apparent solubility of ametryne, atrazine, and diuron in water was shown by TEMPLE and HILTON (1963), using spectrophotometric analysis, to be increased in the presence of surfactants. Among the numerous herbicide-surfactant interactions, the apparent herbicide solubility was increased to varying extents—from almost nothing to as much as 13-fold. Phytotoxicity of the surfactants *per se* to *Cucumis sativus* seedlings was greatest with the cationic type, followed in order by nonionic and anionic. Ability of the three types to enhance the solubility of diuron followed in the same order. The cationics were also most effective in solubilizing the two triazines, although the effective order of the other two types were reversed in the case of these compounds.

The relative potential of anionic and cationic surfactants for herbicidal enhancement and the effects of structural variations in these surfactant types has been comprehensively studied by JANSEN (1965 a). For example, he found that the activity of amitrole, dalapon, 2,4-D, and dinoseb were enhanced in combination with straight chain isomers of an anionic alklbenzenesulfonate surfactant and became progressively greater as the position of the benzene ring was moved toward the center of the 12-carbon *n*-dodecyl alkyl group. Other structural modifications providing marked enhancement included a highly branched dodecyl alkyl group, and substitution of an ester oxygen atom for the benzene ring. Among cationic surfactants, variations in the alkyl structure of ethoxylated alkylamine types affected toxicity of the various herbicides differently, depending upon pH. Although structural variations in this class of surfactant markedly affected certain physical and chemical properties of the spray, such properties could not be correlated consistently with herbicidal effectiveness. It is of interest that in these studies optimal herbicidal enhancement occurred only at the relatively high surfactant concentration of one percent.

Although nonionic surfactants may not necessarily promote greater foliar penetration of the associated toxicant than other types, they are generally desirable because of their greater comptability when formulated in water of high salt content and their reduced ability to interact, sometimes detrimentally, with the toxicant. Since nonionic surfactant molecules contain both hydrophobe and hydrophile moieties, they can be variously synthesized to achieve specific solubility characteristics. One of these types, the alkylphenol polyoxyethylene glycol ethers, has been evaluated for its structure-activity relationships (SMITH *et al.* 1966). This group consists of an alkylphenol hydrophobe and an ethylene oxide (EO) hydrophile. The number of moles of EO in the molecule determine its surface tension and solubility characteristics—more specifically, its hydrophilic-lipophilic balance (HLB). When used in combination with amitrole, dalapon, and paraquat as a foliar spray on maize, the type herbicide, the concentration of the surfactant, and its

EO content all markedly influenced maximum toxicity. Depending upon the surfactant homolog, its concentration, and the herbicide with which it was used, toxicity generally reached a maximum somewhere between eight and 15 moles of EO, above which it gradually decreased. Several exceptions were noted, however. Variations in the hydrophobic portion of the surfactant molecule (octyl-, nonyl-, and laurylphenol types) were of less importance in influencing toxicity. Buchanan and Staniforth (1966) studied the relative toxicities of about 100 surfactants of various ionogenic types, using several bioassays, including ultrastructural effects on leaf tissue of *Elodea* and *Glycine max*. Although exceptions were noted, they generally observed an increase in EO content to be associated with decreased toxicity.

From the above and other studies, it seems apparent that capability of a surfactant to enhance penetration is a function of both HLB and certain other chemical or physical properties. HLB is expressed numerically and ranges from lipophilic values of less than two, through a neutral range of about ten, up to extremely hydrophilic values of 30 or more. These values may be determined by measuring the distribution of the nonionic surfactant between aqueous and isooctane phases, as described by Greenwald *et al.* (1961). Surfactants of relatively high HLB promote normal oil-in-water emulsions, whereas those of low HLB tend to promote water-in-oil or "inverted" emulsions. Such inverts have become increasingly useful in recent years because of their ability to be sprayed with only a minimum formation of fine droplets and resultant minimized tendency to drift. Each pesticide-carrier-target system will usually display an optimum HLB requirement for the surfactant employed. According to Behrens (1964), if different chemical families are tested, the optimum effect is usually observed at the same HLB in all families even though the maximum level of efficiency achieved may vary from family to family. The optimum surfactant HLB requirement will, however, change from system to system, as exemplified by the work of Jansen (1964). In evaluating the effect of the water-soluble triethanolamine salt and of the oil-soluble butoxyethanol ester of 2,4-D on *Glycine max*, he found that a surfactant of relatively high HLB had little influence on the activity of the amine in water, whereas in a nontoxic oil carrier the amine activity proved to be inversely proportional to surfactant concentration. Ester activity in oil was progressively enhanced with increased concentrations of the same surfactant. A lipophilic surfactant of low HLB, on the other hand, did not alter amine activity in either water or oil, but gave greater enhancement of ester activity than did the more hydrophilic surfactant.

The variable and yet intricate relationship of HLB to the pesticide-plant complex is further illustrated by several recent investigations. Thus, Umoessien *et al.* (1967) examined the effect of surfactant HLB on enhancing foliar and root absorption of linuron and prometryne. Although these herbicides differ in lipid and water solubility, absorption and general phytotoxicity of both were enhanced by all of the blended nonionic surfactants tested, within the HLB range of 5.4 to 15. A more positive effect of HLB was noted by SIROIS (1967), who investigated the toxicity of various surfactant-2,4-D combinations to *Lemna minor.* He found a distinct relationship between lipophilicity of the homologous series of nonionic ethylene oxide-ether surfactants tested and both their inherent phytotoxicity and ability to enhance the activity of 2,4-D. In marked contrast, EVANS and ECKERT (1965) observed an *inverse* correlation between lipophilicity of surfactants and their ability to increase the activity of paraquat against *Bromus tectorum.* Of some 22 surfactants tested, 17 reduced yield of dry weight from 73 to 94 percent of that of the controls. Paraquat without a surfactant present reduced yield only 12 percent. MORTON and COMBS (1969), studying the effect of picloram-2,4,5-T (triethylamine salts) on greenhouse and nursery seedlings of three woody species, found greatest herbicidal enhancement with surfactants within the HLB range of 13.3 to 15.4. Surfactants with ether linkages were more effective than those with ester linkages when applied to *Prosopis juliflora,* but both classes were equally effective on *Quercus virginiana* and *Smilax bona-nox.* In an effort to reduce the volume of carrier needed for treatment of *P. juliflora* with 2,4,5-T, we have also examined the influence of several lipophilic surfactants in an all oil system (HULL and SHELLHORN 1967). Seedlings were treated on two basal leaves with the butoxyethanol ester of 2,4,5-T in various combinations of phytotoxic (diesel) and nontoxic oils and nonionic surfactants of HLB's of 8.6 and 1.8. In the majority of combinations the herbicidal activity was not significantly affected by the addition of a surfactant. However, when sorbitan monolaurate (HLB 8.6) was used in combination with diesel oil as a carrier, subsequent apical epinasty and growth repression was greater than with the other combinations.

Some of the apparently diverse effects of surfactants in inducing penetration of different solutes have been mentioned by CURRIER and DYBING (1959). For example, the inability of Tween 20 to enhance ^{32}P penetration into bean leaves, but its potential for increasing the response of maize plants to GA ten-fold. Also, the fact that Triton X-100 may improve phosphorus absorption but depress magnesium absorption. With herbicides, NAKAMURA (1967) observed that the surfactant NK 6601 markedly increased the activity of simazine, atrazine, and atratone against *Digitaria adscendens,* but had little influence on the activity of simetryne and ametryne. It seems probable that at least a portion of these diverse influences may be attributed to chemical interactions in some cases between the surfactant and the active ingredient, particularly when one of the ionic type surfactants is employed. Such an interaction has, in fact, been demonstrated by SMITH and FOY (1967) in mixtures of paraquat and anionic surfactants. By monitoring ultraviolet absorbance of such mixtures, and also the adsorption of ^{14}C-paraquat at solution-air interfaces, they were able to demonstrate significant interactions between surfactant and herbicide. The interactions were further confirmed by the altered activity on maize foliage. The studies of FREED and MONTGOMERY (1958) on foliar absorption of amitrole in bean plants demonstrated the important role of a surfactant in depressing surface tension, but also suggested the relationship of molecular interaction between surfactant and herbicide as perhaps being equal or greater in importance than the lowering of surface tension. The interaction was again noted (HUGHES and FREED 1961) in studies involving the absorption of IAA by the primary leaf of bean plants. The degree of interaction was shown to be markedly influenced by the chemical nature of the surfactant used.

The important capability of surfactants to increase wettability and the spreading of spray droplets should be mentioned. Detailed discussions dealing with associated phenomena, including surface effects, interfacial tension, contact angle of droplets, and also techniques of measuring such phenomena, have been presented by DIMOND (1962) and GOULD (1964). Foliar retention is an important factor, not only in influencing residue and biological enhancement of the specific spray ingredient, but also (in the case of herbicides) of improving selectivity. Retention is a function of both the nature of the foliar surface and that of the spray itself. DAVIES *et al.* (1967), for example, found a high retention of ioxynil on *Sinapis alba* foliage either without a surfactant or with the inclusion of 0.1 or 1.0 percent Tween 20 in the spray solution. However, retention of the herbicide on leaves of barley and pea was obtained only when the surfactant was included, particularly at the higher rate. Without a surfactant present, the *Sinapis* plants retained 26X more ioxynil than did barley plants, but the addition of 0.1 or 1.0 percent Tween 20 increased the retention by barley, thereby reducing the differentials to 11X and 8X, respectively. With the inclusion of four percent Lovo®, an amine stearate, along with a member of wettable powder formulations of both insecticides and fungicides, AMSDEN (1962) obtained markedly increased foliar retention. This was largely attributed to reduced evaporation from the falling droplets, and could be experimentally demonstrated in a large heated spray tower. In addition, the additive resulted in superior rain fastness and improved biological activity which enabled the use of less active ingredient.

The effect of plant species, leaf age, and leaf surface (upper vs. lower) upon absorption has been previously discussed. That the same factors influence wettability and retention seems apparent from the work of ASHWORTH and LLOYD (1961). In cabbage they found that the youngest apical leaves and the oldest basal leaves were more readily wetted than the larger middle leaves. Older plants had a greater number of leaves difficult to wet; also, wilted leaves were more wettable, as were upper surfaces as compared to the lower. The addition of surfactants to the spray should minimize differential wetting of this nature found within a single plant, as well as minimizing the different wetta-bility characteristics among different species. The possibility of adjusting selectivity of herbicidal treatments by this technique alone thus seems quite possible.

It is often important, particularly in experimental work, accurately to measure wetting ability of a solution. In their studies described above ASHWORTH and LLOYD employed a specific length of cotton tape having a standard weave. The surfactant concentration necessary to achieve complete wetting of this tape in 15 seconds accurately determined the wetting capabilities of the surfactant and was related to the wettability obtained when applied to a cabbage leaf. In an attempt to assess the wetting ability of some 65 surfactants on several leaf types FURMIDGE (1965) used several tests to predict the wetting. Methods included visual observation, evaluation of surface properties of the formulations, and air-displacement techniques. None of the tests was completely accurate in predicting wettability unless the appropriate target surface and the effects of spray droplet impact were also taken into consideration, *i.e.*, only by actually spraying the exact formulation on the target could one be certain of the outcome. Contact angle was, however, of value in recognizing extreme cases of complete wetting, and the tape-wetting test proved useful in providing a rough guide for evaluating the wetting capabilities of anionic and nonionic surfactants. Another test of interest, which may be of more use in evaluating the wettability of leaves than estimating the wetting potential of surfactants, has been described by AMSDEN and LEWINS (1966). It involves dipping the entire plant in a 0.5 percent solution of crystal violet in distilled water, and quickly removing it. Such a solution has a surface tension of 60 dynes/cm.^{2} or higher, and the solution remains readily visible on any easily wet areas. The test should be useful in evaluating the effect on wettability of such factors as leaf age, rubbing or abrasion of the foliage, greenhouse vs. field plants, effects of various soil-applied chemicals, etc.

It has for some time been a question as to the extent to which a surfactant molecule itself penetrates the leaf when applied in combination with various active materials which are known to penetrate. Evidence obtained by SMITH and FOY (1966) with bean, barley, and cotton, indicates that the surfactant does indeed enter the leaf, even if only to a limited extent. Using Tween 20, ^{14}C-labeled on either the oxyethylene or the fatty acid portions of the molecule, they found that the latter label especially could hardly be detected outside of the treated area, which included both leaves and roots. Following absorption via either route, limited movement did occur, principally in the phloem in the case of the fatty acid label, and in the xylem for the more hydrophilic oxyethylene-labeled moiety. This, and additional work (FOY and SMITH 1969) demonstrating some transport of the label following application of ^{35}S-sodium lauryl sulfate, suggests that what

little is transported to any considerable extent no longer remains in the
form of the original surfactant molecule but rather becomes part of
other metabolites. It seemed probable that, once having entered the
cuticle, the surfactant molecules may align themselves in monolayers
with their nonpolar ends oriented in the cutin and wax. The protruding
polar ends would thus form a layer, the thickness of which would be
dependent upon the length of the hydrophilic chains of the molecules.
Such layers could act as "hydrophilic channels" which would attract
water and allow for the diffusion of pesticides or other molecules, depending upon their chemical or physical properties. The binding of
surfactant molecules in this manner was suggested by the fact that less
than two to three percent of the activity of the labeled nonionic surfactants was translocated away from the treated leaf.

Another possibility which has been considered in an effort to further enhance herbicide penetration and activity is the incorporation of
the active moiety and a surfactant into a single molecule. Investigating
some 30 such molecules formed of long-chain alkylamine salts of 2,4-D,
JANSEN (1965 b) found that water solutions of 17 of the salts were
herbicidally active on *Glycine max*. Activity was correlated with the
hydrophilic characteristics of the surfactants and surface activity generally followed that of the surfactant moiety of the molecule. Although
it is assumed that the 2,4-D portion of the molecule reached the site
of action as 2,4-D acid, following hydrolysis, the relative motility of
each of the two moieties is not known. Additional work of this nature
would be desirable, perhaps utilizing ^{14}C labeling of the different
molecular components, and in turn comparing their relative motility
with that of the same herbicide and a separate surfactant applied in a
mixture. Data from the present work suggests that herbicides move
into the plant by two distinct pathways—one hydrophilic and the other
lipophilic, and that the relative importance of the two pathways is
dependent not only on the plant species involved, but also on the herbicide formulation and the nature of the carrier. Surface and interfacial
tensions, turbidity, specific conductivity, and pH of the solutions were
not individually correlated with toxicity or, presumably, with enhanced
2,4-D penetration. However, a combination of interfacial tension, turbidity, and conductivity did provide some indication of relative solubility characteristics and expected 2,4-D toxicity. Studies of this nature
provide information on enhancing penetration of certain pesticides,
and also may suggest techniques whereby residues in specific portions
of the plant may be minimized. As pointed out by BEHRENS (1964),
the possible problem of surfactant residues on food or forage crops
must be considered, and compounds selected having minimum residue
characteristics. The trend of encouraging experimental work principally
on compounds previously proven free of mammalian toxicity is desirable and should be continued. Also, the often detrimental effects of
surfactants to soil and water must be kept in mind. A continued search

should be made for surfactants which are biodegradable and of insignificant mammalian toxicity, and yet which are biologically effective in enhancing the activity of the pesticide on its target.

2. Oils.—A straight oil or an oil-water emulsion carrier is known to facilitate foliar absorption of some pesticides under certain conditions. For example, AYA and RIES (1968) found that the inclusion of a paraffinic mineral oil at six percent (v/v) increased the absorption of ^{14}C-amitrole by *Agropyron repens* leaves eight- to 14-fold, as compared to an aqueous carrier. Absorption of amitrole-T was also significantly enhanced, though not to the same extent. In the case of herbicides, a carrier containing at least a small fraction of oil sometimes proves beneficial, particularly for the treatment of woody plants. Evaporation from a falling oil or oil emulsion droplet is slower than from a water droplet of the same size, the difference perhaps assuming greatest importance in aerial sprays carried out under arid conditions. It does seem possible, however, that the supposed advantages of an oil carrier may be partially related to the purely psychological effect of the applicator's observing a rapid development of foliar necrosis, which most oils will induce. If herbicidal effect is determined as a function of percentage kill of the woody plant as measured about two years after treatment, rather than from relative defoliation after only several months, probably the major part of experimental work will show no significant enhancement of herbicidal action from the addition of oil to the spray formulation. This statement may be truer of some herbicides than of others and is, of course, dependent upon an optimum quantity of surfactant being present in the purely aqueous carrier.

In an attempt to improve control of annual grasses in maize and also to reduce atrazine rates and overcome residue problems, BANDEEN and VERSTRAETE (1967) have evaluated emulsified oils and surfactants as adjuvants for this herbicide. Diesel oils with aromatic contents as high as 25 percent were quite safe at 1.5 gallons/acre (gal./A) but did not give maximum weed control, perhaps because of their low viscosity and rapid evaporation. Oils having a viscosity of over 70 and unsulfonatable residues of over 75 percent were preferable and often effective at only one gal./A. Also evaluating atrazine for weed control in maize, BLACK (1968) found the ratio of oil to surfactant to be of particular importance for formulations in an aqueous spray base. A blend consisting of four parts oil to one part Atplus 300 proved considerably more effective than a standard emulsifiable oil, allowing the volume of the adjuvant to be reduced from one gal./A to about ¼ of this quantity.

Most work suggests that oils penetrate principally via the stomata or through minute fissures in the cuticle. They may move in intercellular spaces, or remain therein for extended periods of time, depending upon toxicity. Many pesticides apparently increase the permeability of plasmatic membranes through an interaction involving both inter-ionic forces and van der Waals attraction between the membrane and the nonpolar portion of the penetrating molecule. According to BRIAN (1964), phytotoxic oils may exert a similar type of action because of their affinity for the membrane due to van der Waals forces. The resulting denaturation and disruption of the membrane increase its permeability to the extent that leakage of cell sap and osmotic changes soon induce death.

As in the case of surfactants, the relative phytotoxicity of the oil appears to have some influence on its ability to enhance absorption of the associated pesticide. There is some indication that a certain degree of toxicity is desirable, although available data are not all in agreement on this subject. LEVI (1960) found that oils of low toxicity increased the herbicidal effectiveness of 2,4-D on *Typha angustifolia*. Highly toxic oils were able to enhance cuticular penetration of the herbicide, but this action was apparently more than counteracted by the rapid killing of conductive tissue and resultant impeded translocation. In studies with greenhouse seedlings of *Prosopis juliflora,* we likewise found that absorption and translocation of 2,4,5-T was considerably greater when carried in a nontoxic oil-water emulsion as compared to a diesel oil emulsion (HULL 1956). Subsequent treatment of field plants by aerial spray, however, failed to disclose any significant effect of type-of-oil. Plant species apparently differ considerably as to their relative resistance to petroleum oils. For example, by means of neutral red uptake and plasmolysis in calcium chloride solutions, MINSHALL (1961) determined the time required for a ten percent solution of tetrahydronaphthalene in paraffinic oil to kill 50 percent of the ground parenchyma cells in petioles of different field-grown plants. The umbelliferous species *Daucus carota* and *Pastinaca sativa* showed greater tolerance to petroleum oils than did *Brassica kaber* or *Beta vulgaris,* apparently due to the greater resistance of their cellular membranes. Differences in time required to establish the LD_{50} in field plants as compared to greenhouse plants existed in some species but not in others.

That herbicidal response may be enhanced by the use of various types of oils, and that such enhancement is markedly variable among different species, has been clearly demonstrated by JANSEN (1966). He examined the postemergence toxicity of wettable powder formulations of atrazine and diuron to nine weed and crop species, when the carrier was made up of water or of a five or ten percent emulsion of various oils in water. Oils were selected with respect to their petroleum source (naphthenic or paraffinic) and viscosity. Responses to the two herbicides varied among the different species, but the phytoxicity resulting from low rates of either compound was differentially increased by the addition of oils or surfactants. The specific toxicity spectrum for each herbicide, with the phytobland oils, was interrelated with chemical nature and properties of the oil, phase volume ratio of oil to water, and spray volume. The responses suggested distinct possibilities of increasing selectivity and minimizing residue hazard by the careful selection of type and concentration of oil for each specific weed-crop-herbicide situation.

3. Humectants.—Foliar penetration of organic solutes is sometimes improved markedly upon the addition of humectant, although different types of pesticides, growth regulators, etc., seem to differ considerably in their potential for such enhanced penetration. Perhaps one reason for the apparent variability in this respect is simply because of the paucity of data available for some classes of compounds. In spite of experimental work which has demonstrated humectant-enhanced penetration of certain compounds, humectants do not seem to have gained wide acceptance in practical application of pesticides and related materials, such as has been obtained by surfactants and oils.

Probably the greatest degree of humectant-enhanced penetration has been shown in the case of antibiotics. GOODMAN (1962) demonstrated improved foliar uptake of streptomycin and resultant disease control with inclusion of methyl cellosolve in the formulation. By perforating the waxy component of the cuticle the cellosolve was apparently able to increase penetrability of the antibiotic. He also reviews the work of GRAY (1956) which, briefly, demonstrates increased foliar uptake of streptomycin and other antibiotics by as much as 55-fold upon addition of one or two percent of various glycols, sugars, surfactants, and particularly of glycerol, to the spray formulation. It is of interest that surfactants were generally without effect and could even be antagonistic toward the action of the glycerol. In discussing both foliar and root uptake of systemic fungicides and bactericides, BRIAN (1967) also mentions the enhanced absorption achieved with certain humectants, as well as the fact that some plant hormones appear to speed up entry of these systemic materials.

Dust mixtures of various plant growth regulators, as mentioned by MITCHELL et al. (1960), have shown increased effectiveness when formulated with certain hygroscopic agents. In other instances such agents seem to be without effect, as found by BISWAS and ROGERS (1963) in their studies of GA absorption by *Matthiola incana*. The physiological responses resulting from foliage sprays of this compound were not significantly altered upon addition of one percent glycerol or several different glycols, suggesting no effect upon absorption.

Some of the earlier work relating to enhancement of different herbicides by such humectants as glycerol, propylene glycol, and molasses has been reviewed by CURRIER (1959). The relative benefit obtained from the use of humectants in combination with herbicides is apparently more variable than is the case with other classes of pesticides and regulators. Enhanced herbicidal activity may of course be a function of foliar retention as well as increased absorption. Both factors may be affected by humectants. In studies involving the effect of 2,4-D on *Glycine max*, ENNIS et al. (1952) observed increased herbicidal inhibition upon the addition of 0.5 percent Carbowax® 1500 to the aqueous formulation and attributed the action to improved foliar retention. HUGHES (1968) investigated the effects of glycerol, propylene glycol, and polypropylenediol, each at one, five, and ten percent, when used as adjuvants to amine salts of 2,4-D, 2,4,5-T, and combinations thereof. The foliar injury resulting to greenhouse specimens of *Tamarix pentandra* was increased to the greatest extent with polypropylenediol, particularly at the median concentration. In contrast to these studies PRASAD *et al.* (1962) found that penetration and translocation of dalapon by bean and barley plants was unaltered upon addition of 0.1 to 5.0 percent glycerol to the aqueous formulation, although the nonionic surfactant X-77 at 0.1 percent did enhance activity. We have examined the effect of several humectants on *Prosopis juliflora* seedlings when used in combination with different phenoxy herbicides. Although borderline enhancement was observed in a few such instances, none of the formulations containing humectants showed a significant herbicidal enhancement when applied to large field specimens.

4. **Film-forming substances and deposit builders.**—A limited amount of research with film-forming compounds has been carried out in an effort to enhance pesticide absorption and prevent its leaching, as well as to minimize transpiration or reduce the foliar absorption of toxic quantities of salt. The advantages of such compounds have been comprehensively discussed by EBELING (1963), as has the influence of adhesives and related substances. Results have been variable—sometimes disappointing. Investigating the fungitoxicity of copper oxychloride sprays on greenhouse specimens of *Vicia faba*, EVANS *et al.* (1966) did, however, observe a generally enhanced effect upon the addition of surfactants. Such agents did not materially affect the rain-resistance of the surface deposits and, when they were further combined with polymeric materials (emulsions of polyvinyl acetate, polyvinyl chloride, or polyvinyl butral), an improved biological performance was obtained. The polymerics improved the tenacity of the copper oxychloride without reducing its fungitoxicity. In additional field experiments on control of potato blight (*Phytophthora infestans*), the advantages of enhanced tenacity and good distribution from the polymeric emulsions unfortunately did not materialize. Work at this laboratory (HULL 1964 a) demonstrated that the addition of a latex suspension to a 2,4,5-T ester in a ten percent emulsion of nontoxic oil in water resulted in greatly increased apical epinasty following application to the basal leaves of *Prosopis juliflora* seedlings. Mean epinasty 72 hours after treatment was 59°, whereas epinasty resulting from all other combinations of some four 2,4,5-T formulations and five adjuvants ranged only from 8° to 16°. The formulations were subsequently tested on large field trees, but the latex combination did not cause an increased percentage of kill. Additional work with seedlings of this species (HULL 1964 b) indicated that when the 2,4,5-T ester was simply emulsified in water, no additional activity was obtained upon addition of latex. However, when

glycerol was included in the above combination, in concentrations ranging up to 25 percent of the total formulation, subsequent repression of root growth was directly proportional to glycerol concentration, suggesting an enhanced foliar penetration and possibly basipetal translocation.

Another possible use of film-forming substances is related to their ability to form weather-resistant films for retaining chemical repellents. In reforestation programs it is often necessary to use some type of repellent to protect seedlings from damage caused by various species of mammals. BULLARD and CAMPBELL (1968) investigated numerous adhesives for their ability to maintain mammalian repellents on seedlings of *Pseudotsuga menziesii*. Of those which were easily formulated and not readily weathered from the foliage, none was phytotoxic in tests lasting from three months to one year. Most effective, as measured by permanency of the film and its ability to hold incorporated sand, were certain of the acrylic-based emulsions. With a somewhat different objective in mind, MALCOLM *et al.* (1968) evaluated different leaf coatings for their ability to reduce uptake of chloride from brackish water used for sprinkler irrigation. Significant reductions in foliar uptake of chloride occurred during sprinkling with a sodium chloride solution when leaves were coated with acrylic polymer latex. Silicones and other latices either increased uptake or were not effective in reducing it. In an attempt to imitate a leaf coating in an inert system, the investigators used filters having a maximum pore diameter of $0.20\ \mu$ and impregnated them with a range of concentrations of latex materials. The diffusion resistance of the filters to carbon dioxide, oxygen, and water vapor was subsequently followed. Increasing the latex concentration from ten to 25 percent caused only a relatively small increase in diffusion resistance to the gases, but both 50 and 75 percent concentrations markedly increased diffusion resistance. With the objective of reducing foliar transpiration by means of a thin plastic coating applied to the leaf surface, WOOLLEY (1967) investigated the relative diffusion resistance of ten different types of plastic films (in an artificial system) to carbon dioxide and water vapor. In order to permit photosynthesis, such a film on the leaf would have to be more permeable to carbon dioxide than to water. However, none of the films tested had a carbon dioxide permeability as great as its permeability to water.

Various deposit builders, also called thickening agents or particulating agents, are available for the purpose of minimizing the proportion of very small spray droplets which may otherwise cause a drift hazard in certain pesticide applications. Such compounds are generally very effective for this purpose. Particulating agents may consist of hydroxyethyl cellulose, polysaccharide gums, or various other polymeric materials which swell in water, producing a thick mixture of somewhat granular appearance. Ordinarily no claim is made that they may enhance activity of the pesticide with which they are used. Most work with these compounds has suggested that they indeed do not generally alter such activity significantly, one way or the other. One notable exception, however, has just been reported. HAAS and LEHMAN (1969) applied an aerial spray of picloram to a mixed hardwood forest, at the rate of one lb./A. The formulation was made up in an aqueous carrier with and without the addition of hydroxyethyl cellulose (HEC), a water swellable plastic polymer. With most species the percentage of plants dead 15 months after treatment was significantly greater when HEC was included. With the overstory species the percentage kill (with and without HEC, respectively) was as follows: *Quercus stellata* 53 and 13, *Q. marilandica* 65 and 15, *Ulmus alata* 33 and 15, and *Carya texana* 23 and 36. The understory species ran as follows: *Ilex vomitoria* 12 and 0, *Vaccinium arboreum* 31 and 24, *Callicarpa americana* 45 and 10, and *Crataegus marshallii* 73 and 12.

That deposit builders, used in combination with oil-base sprays, may inhibit foliar penetration to some extent is suggested by the work of GUNTHER *et al.* (1946). Investigating surface residues of DDT on citrus foliage following spraying for the control of California red scale (*Aonidiella aurantii*), they found that both the insecticide and kerosene carrier penetrated into the leaf tissues almost immediately, only to be reissued to its surface over the next 24 hours. Evaporation of the solvent subsequently left a surface deposit of DDT. When aluminum stearate was added to the formulation as a petroleum gelling agent, it largely prevented this immediate penetration into the inner tissues, resulting in a much greater initial deposit of DDT; however, after 24 hours, the surface residue resulting from the spray formulations with and without aluminum stearate were very nearly the same.

From the limited number of and rather diverse responses described above, it seems evident that considerably more research with film-forming and particulating agents is needed before general recommendations can be given. One thing which should be kept in mind for spray applications of these materials is the usual necessity of sufficiently large nozzle orifices and ample spray pressure, such as may be achieved with gear-driven pumps. Nozzle clogging may otherwise be a problem.

5. **Growth regulators.**—Enhanced penetration of several organic and inorganic substances, as well as their movement within the leaf, has been experimentally induced with several classes of growth regulators. Any advantages gained for practical field application of pesticides, however, have apparently been rather slight, or at least have not gained wide acceptance. With kinins, MOTHES and ENGELBRECHT (1961) found that when excised leaves of *Nicotiana rustica* were treated on the right upper quarter, and ^{14}C-glycine was applied four days later as a drop to the same side, but proximal to the kinetin-treated area, subsequent movement of the label was preferentially directed toward the kinetin-treated area. When leaves were maintained in the dark, ^{14}C activity was found in several amino acids (primarily glycine and serine)

and in carbohydrates. Leaves kept dark for long periods, or starved, or the white parts of variegated leaves, often failed to show the effect, suggesting the necessity of ATP for the process. MÜLLER and LEOPOLD (1966 a) established that movement of ^{32}P was preferentially toward the base of excised maize leaves, indicating a natural mobilization center in that area. They also found that localized application of kinetin to a leaf caused an acceleration of ^{32}P movement toward the kinetin center, and an increased accumulation of the isotope at the center. The attracting activity of such centers was additionally enhanced by increasing amounts of materials already accumulated—a sort of "self amplifying" process. They further observed (1966 b) that when two mobilizing centers were separated from one another, they competed for a coherent transport system, the direction and velocity of which was determined by the balance of the mobilizing forces. Unlike ^{32}P, the kinetin-treated centers did not accumulate ^{22}Na, ^{86}Rb, ^{36}Cl, or ^{131}I. The pattern of stimulated ^{32}P movement in these experiments was strongly suggestive of a mass flow transport in the phloem. The above work is reminiscent of the finding of OSBORNE and HALLAWAY (1961), that when spot applications of 2,4-D were made to attached *Euonymus japonica* leaves, the treated areas remained green while the surrounding tissues yellowed. Labeled carbon compounds, following exposure of the leaf to $^{14}CO_2$, either moved into or were preferentially retained in the treated areas, which continued to respire normally in contrast to declining respiration in the chlorotic areas. The treated cells apparently acted as metabolic sinks to which nitrogen and carbon materials were drawn from surrounding cells, resulting in premature senescence in untreated leaf parts which contained only a relatively low concentration of the applied 2,4-D.

An additional characteristic of certain kinins, as observed by LUKE and FREEMAN (1968), is their facility for increasing absorption of victorin. This pathotoxin is produced by *Helminthosporium victoriae*, and causes Victoria blight of oats. One kinin, N^6-2-butoxyethylaminopurine, increased the total liquid absorption when applied with victorin to excised leaves, and also decreased the time required for wilting. The action was apparently one of increased victorin uptake, rather than alteration of its toxicity, and was associated with stimulated transpiration.

Other types of growth regulators and hormones apparently capable of speeding up the foliar penetration of systemic fungicides and bactericides have been discussed by P. W. BRIAN (1967). The uptake of ^{14}C-propazine by detached leaves of *Camellia sinensis* floated on an agar solution containing 20 p.p.m. of this herbicide was found by BISWAS and HEMPHILL (1965) to be differently affected by two growth regulators. With no regulator present, upper leaf surfaces absorbed propazine to the extent of 3.5 p.p.m., compared with 15 p.p.m. for the lower surfaces. Addition of GA_3 increased the quantity of propazine absorbed by upper surfaces to 9.5 p.p.m., while that absorbed by lower surfaces was not significantly altered. IAA, on the other hand, increased absorption by both surfaces, the upper proportionately more than the lower. Measurement of propazine diffusion through cuticles isolated from upper and lower surfaces actually showed a slight decrease upon addition of either IAA or GA, suggesting that the enhancement is elicited by an extra-cuticular mechanism. Uptake induced by growth regulators may also be a function of concentration of the regulator. For example, MASUDA (1953) noticed that the permeability of epidermal cells of onion bulb scales to urea and glycerol was increased in the presence of $10^{-8}M$ IAA, while $10^{-3}M$ reduced the permeability. Foliar uptake of inorganic salts is also influenced by different growth substances when the latter are absorbed via the roots (HALEVY and WITTWER 1965). They found that uptake of ^{86}Rb, applied as RbCl to leaves of bean plants, was increased by prior application of either GA_3 or NAA to the culture solution. CCC and B-9 had no effect on uptake, whereas both BA and Phosphon inhibited it. The fact that subsequent effects of these regulators on translocation proved to be quite different from the absorption effects suggests the complete independence of the two processes.

With herbicide treatments there is some indication that the uptake and subsequent activity of such compounds may be significantly enhanced through prior or simultaneous application of either the same or a different herbicide. CHOW et al. (1966) found that ^{14}C-silvex applied to pads of Opuntia polyacantha was absorbed to the extent of five percent of the quantity applied over a six-day period. In pads pretreated 24 hours previously with a four percent aqueous solution of dalapon, silvex absorption was increased to eight percent. Evaluating foliar uptake of picloram by a gas chromatographic technique, DAVIS et al. (1968 b) demonstrated a significantly increased 72-hour absorption by the small leathery leaves of Ilex vomitoria when paraquat was also included in the formulation. Uptake of picloram over 20 hours in Prosopis juliflora was similarly increased in the presence of 2,4,5-T; however, 2,4,5-T uptake was inhibited by picloram. The comparatively delicate leaflets of the latter species are perhaps more sensitive to toxicants than the Ilex leaves, in which event too rapid a contact injury may result in inhibited uptake of another compound. During a study of the effect of pH, temperature, and other factors on uptake of ^{14}C-dicamba by leaf sections of Triticum aestivum and Polygonum convolvulus, QUIMBY and NALEWAJA (1968) found that when a 100-fold concentration of unlabeled dicamba was added to the bathing solution along with the ^{14}C-dicamba, there was an apparent increased uptake of the labeled compound. Such a facilitated absorption of dicamba anions is not unlike the "self amplifying" process described above by MÜLLER and LEOPOLD (1966 a) for ^{32}P. When the leaf sections were boiled the quantity of ^{14}C-dicamba was about doubled, suggesting that heat denaturation increased the number of binding sites on adsorptive protein.

6. DMSO and other penetrants.—Although DMSO was first synthesized in Germany in 1867, its remarkable ability to penetrate mammalian and plant tissues deeply was recognized relatively recently. Its additional capacity to enhance penetration of substances with which it is mixed has drawn the interest of workers in both medical and agricultural fields. The biological actions induced by this compound have been comprehensively reviewed by LEAKE (1967). Studies carried out by GARREN (1967) with dilute solutions of ^{35}S-DMSO demonstrated that it quickly becomes systemic in young pear trees following application to the bark tissues. He also noted that when $Na_2H^{32}PO_4$ was applied to the soil or culture solution in which strawberry plants were growing, the quantity of ^{32}P reaching the leaves was approximately doubled if a dilute DMSO solution had been previously added to the substrate. SCHMID (1968) also observed that DMSO in concentrations of up to ten percent stimulated the uptake of ^{65}Zn by excised barley roots. However, at the same concentration range it strongly depressed uptake of ^{22}Na and ^{86}Rb. It apparently did not alter membrane permeability, since roots treated with desorption solutions containing ten percent DMSO did not lose more of the preferred ion than did roots desorbed in straight aqueous solutions. DMSO did, however, reduce oxygen utilization, suggesting that it may be a poisoning agent which interferes with cation transport by attacking some aspect of metabolism rather than by altering membrane permeability. MUSSELL *et al.* (1965) obtained occasional enhancement of activity from 2,4-D when it was applied with DMSO as the carrier. However, following observations on plant membrane properties before, during, and after treatment with DMSO, they could detect no significant changes and likewise could not attribute the DMSO effect to altered membrane behavior. Rather, the above and additional observations suggested that the effects on solute penetration were due to the physical characteristics of DMSO itself.

There is some evidence that higher concentrations of DMSO may be injurious to cytoplasmic membranes of plant cells. RICHTER (1968) studied the resistance of epidermal cells of *Campanula* species to three substances which have been used as cryoprotective agents for mammalian cells, namely DMSO, glycerol, and ethylene glycol. At concentrations of 50 and 100 percent, only glycerol proved to be without toxicity over a one-day period, as indicated by renewed protoplasmic streaming upon removal from the solution and by staining and plasmolytic behavior. DMSO and ethylene glycol, apparently because of their solubility characteristics, both caused injury to the cytoplasmic membranes. The effect of DMSO upon diffusion of glucose through cellulose acetate membranes has also been examined by BLUMBERG and ILAGAN (1968). The DMSO/water solvent had a maximum viscosity at 0.35 mole fraction DMSO, and the permeability constant was found to decrease with an increase in DMSO content and solvent viscosity up to that point. At higher DMSO concentrations of about 0.4 mole fraction and over, the membranes disintegrated. The inhibitory influence of DMSO on glucose transport may thus be related to both viscosity and membrane-damaging effects.

Some of the additional reported physiological effects of DMSO on plants include those of germination, growth, mutation frequency, alkaloid content, and enzymatic activity. SIDDIQ et al. (1968) found that the percentage germination of unhulled rice seeds, as well as survival, was enhanced if they were treated with 2.5 percent DMSO following an eight-hour presoaking in water. Subsequent growth rate, however, was reduced. DMSO concentrations up to 30 percent proportionately reduced germination, survival, and growth rate, and also potentiated the toxicity of ethyl methane sulfonate, as measured by the same responses in the M_1 generation. DMSO concentrations of only 1.4×10^{-3} M (ca. 0.01 percent) in the nutrient culture solution were found by NETHERY and HURTT (1967) to decrease plant height slightly, although this level was essentially nontoxic over a seven-day period. Concentrations of 2.8×10^{-2} M or over significantly reduced fresh and dry weights of shoots and roots. Visual foliar symptoms and autoradiography following both foliar and root application of ^{14}C-DMSO suggested xylem transport.

Significant increases in height growth following a series of foliar sprays of DMSO have been reported by SCIUCHETTI and HUTCHISON (1966). Both height growth and stem dry weight of *Datura tatula* were increased by sprays of two percent (v/v) DMSO, although such treatments significantly decreased root dry weight. This treatment caused a decrease in chlorophylls a and b, and generally increased the alkaloid content to a varying extent in different portions of the plant. Sprays of 100 p.p.m. GA markedly increased height growth of this species and, when combined with DMSO, the effects were additive. Additional work with four species of *Datura* (SCIUCHETTI 1967) showed generally the same response pattern from DMSO, although some differences in effect on alkaloids were apparently species-related. Concentrations of five and ten percent caused varying degrees of leaf burning. The effects of additional growth regulators, including Phosphon and B995, were either additive or potentiated to a certain extent when used in combination with DMSO. The studies of CHANG and SIMON (1968) demonstrated that growth of several types of mammalian cells and of *Escherichia coli* (including the bacteriophage T4 therein) is variously inhibited by DMSO concentrations ranging from two to 12 percent. They also observed a slightly increased ribonuclease and β-galactosidase activity and markedly decreased glycerol dehydrogenase activity. They concluded that the *in vivo* effects of DMSO were principally due to its ability to alter enzyme reaction rates.

As to the ability of DMSO to enhance the penetration of certain pesticides and consequently increase their biological activity, various degrees of success have been reported ranging from no effect to rather extreme effects. Concentration of DMSO seems to be an important factor in this relationship. TSCHIRLEY (1968) cites examples whereby the herbicidal effect of two oz./A of 2,4,5-T on *Acacia farnesiana* and *Prosopis juliflora* increased with increasing concentrations of DMSO. In other experiments defoliation at two weeks and two months following treatment of *Quercus virginiana* and *Rosa bracteata* with varying rates of 2,4-D, 2,4,5-T, paraquat, and picloram was not significantly altered with the inclusion of 0.1, 1.0, or 10.0 percent (v/v) of DMSO in the formulation. BRADY and PEEVY (1968) similarly reported that the same herbicides, as well as dicamba and dichloroprop, did not produce increased top-kill of *Liquidambar styraciflua* and *Quercus marilandica* when DMSO was added at one or ten percent (v/v) or even up to 50 percent in the case of several of the herbicides. In studies with young *Carex cherokeensis* plants BURNS *et al.* (1967) found that absorption and subsequent transport of ^{14}C-2,4-D following application to the third leaf was not affected by two percent DMSO.

A few instances have been reported in which relatively low DMSO concentrations have enhanced uptake of the active ingredient of the spray and its subsequent physiological action. KEIL (1967) found that oxytetracycline controlled bacterial spot (*Xanthomonas pruni*) on peach trees, but was able to obtain a significantly greater control by adding 0.25 or 0.5 percent DMSO to the spray solution. The effect was apparently due to increased penetration and transport of the antibiotic. At harvest the fruit showed no antibiotic residue or off-flavor after seven spray applications, the last 43 days before harvest. KEIL (1965) has also discussed the augmentation of bactericidal or fungicidal activity of organic mercurials, quaternary ammonium compounds, dodine, hexachlorophene, dinocap, and zinc sulfate, which has been achieved with one percent DMSO. In studies on the foliar nutrition of citrus, LEONARD (1967) evaluated numerous aqueous formulations of iron either by spraying or by dipping foliage which had developed severe iron chlorosis. Iron uptake, as determined by the rate and extent of greening, was markedly enhanced with the addition of two percent DMSO. As to herbicides, LAPHAM (1966) found that although DMSO at only 0.5 to 1.0 percent was not toxic by itself, it did increase the activity of 2,4-D on *Alternanthera philoxeroides*, and also the activity of MSMA and DSMA on *Sorghum halepense* and *Verbena* spp. The effect of amitrole on *S. halepense* was greatly increased, but most striking was the combination of DMSO with 2,4-D-picloram mixtures, in which numerous broadleaved species were killed to the ground with no regrowth occurring nine months later. Although 2,4-D at two lb./A in an aqueous carrier would kill no *Verbascum thapsus*, PENNINGTON and ERICKSON (1966) noted that the same rate in a solution of 2.4 percent (w/w) DMSO gave a 62 percent kill of the plants. Dicamba proved more effective than 2,4-D on this species, and its activity was considerably enhanced with even lower DMSO concentrations. DMSO at a volume concentration of one percent was found by BOVEY and MILLER

(1968) to be particularly effective in increasing the desiccation capabilities of paraquat. The somewhat more rapid necrosis observed on white as compared to green leaves of *Hibiscus rosa sinensis* was apparently not a function of absorption rate, but was related to chlorophyll content of the tissue.

Although DMSO at higher concentrations, *i.e.*, much above five percent, is quite phytotoxic in itself, it is with such concentrations that some of its more dramatic effects have been demonstrated, particularly in combination with herbicides. Thus, HART and HURTT (1967) found that 30 percent DMSO markedly enhanced the toxicity of diquat and paraquat to bean plants, either with or without 0.5 percent Tween 20. The same concentration of DMSO augmented the activity of picloram, dicamba, and 2,4-D to a lesser extent, depending upon the presence or absence of a surfactant. The only herbicide the activity of which was not increased at least to some extent with DMSO was 2,3,6-TBA. DMSO-enhanced absorption of 2,4-D was quantified in a somewhat different manner by MUSSELL (1967). He observed that when the acid or lithium salt of this herbicide was dissolved in DMSO, as compared to water, and applied as a droplet to the primary leaf of bean, it was four to five times more effective in inducing abscission of the treated leaf and the leaves above. When the 2,4-D was injected into the hypocotyl, identical symptoms developed regardless of the solvent used, suggesting that the DMSO-enhancement of 2,4-D activity resulted from its effect on penetration rather than on translocation or metabolism. The *n*-octyl ester of 2,4-D emulsified in water was as effective as the acid and salt in DMSO, but the activity of the ester was not further enhanced when carried in DMSO. In experiments with *Prosopis juliflora* seedlings we have noted a similar behavior: the foliar absorption of water-soluble formulations of 2,4,5-T is considerably facilitated in the presence of DMSO but the inherently superior absorptive characteristics of ester formulations are not significantly altered (HULL and SHELLHORN 1968 and unpublished data). Working with the same species we recently observed (SHELLHORN and HULL 1969) that foliar penetration of a mixture of the triethylamine salt of 2,4,5-T and potassium salt of picloram is substantially increased if formulated in a carrier consisting of DMSO, glycerol (or ethylene glycol), nontoxic oil, and water, in the respective volume ratios of 50:25:15:10. Application of 20 μl. of this formulation to the upper surface of each of the two basal leaves resulted in a larger percentage of plants killed than we have obtained with any other combination, including the same herbicides in an aqueous solution of 50 or 100 percent DMSO. The formulation was most effective when emulsified with 0.5 percent of a slightly lipophilic (HLB 8.6) surfactant.

The rather intricate relationship between the herbicide-potentiating ability of DMSO and the concentration used was perhaps first noticed by NORRIS and FREED (1963). Following treatment of one leaf of *Acer macrophyllum* seedlings with ^{14}C-2,4,5-T (triethyanolamine salt) carried in 20 percent DMSO, they could detect no increase in absorption over that obtained in a water carrier. However, carriers of 50 and 100 percent DMSO resulted in absorptive increases of approximately five- and ten-fold, respectively, over that obtained with the water carrier. Subsequent translocation to the roots, on the other hand, was not significantly altered. A somewhat similar behavior of DMSO has also been observed on both growth chamber and greenhouse grown *Prosopis juliflora* seedlings in this laboratory (HULL 1965). The activity of dicamba, of picloram, and particularly of 2,4,5-T (triethylamine salt) applied at the rate of 20 μl. to the basal leaf only of each plant was significantly enhanced if carried in 50 or 100 percent DMSO. In marked contrast, a carrier of 20 percent DMSO was antagonistic toward the action of these herbicides, especially the phenoxy compound.

With dalapon and diuron, BAYER and DREVER (1964) found that activity toward greenhouse grown oats was not significantly enhanced with a carrier of one, ten, or 100 percent DMSO, with one exception. When diuron was formulated with 100 percent DMSO plus one percent of X-77, a nonionic surfactant, the resultant repression of fresh weight increment was markedly greater than with any of the other herbicidal combinations.

The principal mechanism by which DMSO intensifies the penetrability of many diverse molecules has not been clearly identified. The probable destruction of plasmatic membranes above a certain threshold concentration, in itself, could not account for all of the anomalous responses noted above. KEIL (1965) suggests that DMSO may promote transport of the unchanged toxic moiety, but may additionally cause decomposition of certain molecules to toxic substances which may in turn be transported. An inherently nontoxic ingredient could, after DMSO-induced absorption, conceivably alter plant metabolism to such an extent that new biologically active substances are produced. Because of the relatively high chemical reactivity of DMSO itself, it also seems probable that new compounds may be formed when certain chemicals are dissolved therein. Whether some of these reaction products or altered plant metabolites might form toxic residues on crop plants is not currently known. Even if some degree of mammalian toxicity were demonstrated when used in combination with specific pesticides, DMSO may still serve as a useful spray adjuvant in non-crop situations.

In addition to DMSO there are certain other organic solvents which have exceptionally strong penetrability characteristics. Preliminary experiments at this laboratory have shown that the lactam, M-Pyrol®, and the lactone, BLO®, may under certain circumstances be even more effective than DMSO. These penetrants, additionally, are highly phytotoxic to the extent that they could virtually be considered herbicides themselves.

7. **Inorganic salts.**—There is an indication, particularly with some growth regulators and herbicides, that the inclusion of certain inorganic salts in the formulation may increase uptake and biological activity of the active ingredient. For the most part, the degree of enhancement so obtained in experimental work has been of borderline significance, and salts as an adjuvant for spray solutions under field conditions have not been widely adopted. Occasional incompatabilities with the surfactant or active ingredient may be one reason for lack of such adoption. It is also of interest that enhanced uptake of certain sugars is known to occur, particularly from the action of boron. For example, during a study of foliar uptake and transport of ^{14}C from labeled sucrose and carbon dioxide by bean cuttings, WEISER et al. (1964) found that inclusion of boric acid or aluminum chloride in the culture solution significantly increased sugar uptake. No effect on translocation was observed, however, as had been previously reported. Although the essential role of boron in plants is not well known, the increased boron content of the tissue in the boron-fed plants apparently altered metabolism of citric and isocitric acid. Enhanced foliar uptake of sugars may also be induced by boron when added directly to the sugar solution, as noted by NELSON and GORHAM (1957) with ^{14}C-glucose. Uptake of a measured quantity from the surface of a primary leaf of *Glycine max* was only 60 percent complete in three hours; however, when 0.8 percent of sodium lauryl sulphate was included, it was 100 percent complete in 1.5 hours. With an additional five p.p.m. of boron total absorption was achieved in only 15 minutes, although the boron alone caused no significant enhancement of glucose uptake.

Ammonium salts are known to enhance the uptake of NAA. HORS-FALL and MOORE (1962) determined uptake of the sodium salt of this regulator as a function of leaf curvature and, of numerous ammonium salts investigated, found that only those of relatively strong acids actually enhanced uptake. Thus, ammonium sulfate was effective when applied for several consecutive days prior to treatment with the NAA, as well as when applied together in a single solution. The enhancing effect of the salts could not be assigned to the NH_4 moiety of the molecule alone, but apparently was due to a synergistic cation-anion combination between the $=N=$ configuration in NH_4 and the anion from a strongly ionized precursor acid. An interesting effect of ammonium thiocyanate when used in combination with amitrole has likewise been noted. VAN DER ZWEEP (1965) observed that the addition of increasing amounts of this thiocyanate to an amitrole solution for foliar application to bean seedlings resulted in decreased contact toxicity but an increased shoot inhibition. When ^{14}C-amitrole was used in the formulation, dispersal of the label within the treated leaf was increased with increasing concentrations of the thiocyanate. Addition of BA to the amitrole produced an even greater synergistic effect than that obtained with thiocyanate. In contrast to these results with bean, DONNALLEY and Ries (1964) found that in *Agropyron repens*, regardless of time of application of the ^{14}C-amitrole, addition of ammonium thiocyanate to the leaf did not alter the amount of amitrole absorbed. It did, however, greatly increase the amount of ^{14}C translocated.

Several investigators have demonstrated a salt-enhanced selective activity of phenoxy herbicides on broadleaved weeds in monocotyledonous crops such as maize. For example, Ladonin (1961) and Likholat (1962) found that 15 kg./ha. of ammonium sulfate in common application with 2,4-D at a rate af about one to two kg./ha. significantly increased toxicity of the herbicide toward the weeds, and resulted in their effective control. At the same time, the yield of grain was increased by approximately 5.8 metric centners/ha. Al'tergot and Kiselev (1963) investigated the ability of numerous inorganic salts to increase the toxicity of 2,4-D toward various broad-leaved weeds, and found that ammonium nitrate, ammonium sulfate, ammonium acid phosphate, potassium nitrate, calcium nitrate, and calcium dihydrogen phosphate all possessed this ability to varying degrees. Of these salts, ammonium nitrate was most effective. Used at a concentration of ten percent in combination with 0.5 percent 2,4-D, it possessed a toxicity comparable to that of 2,4-D used alone at one to 1.5 percent. Salt apparently enhanced 2,4-D uptake in both monocots and dicots and resulted in distinct changes in enzymatic activity, oxidative processes, and other physiological functions. Salt-induced enhancement of phenoxy herbicide activity with woody plants has apparently been more difficult to demonstrate. Tschirley (1968) applied several formulations of 2,4-D, 2,4,5-T, and MCPA as a foliar spray at rates of 0.5 to two lb./A to five woody species. Evaluation of leaf kill up to four weeks following treatment, and stem kill after one year, indicated that inclusion of ammonium thiocyanate in the formulation at one part to 20 parts of herbicide (w/w) did not significantly alter herbicidal activity, with one possible exception: stem kill of *Rosa bracteata* was 50 percent with thiocyanate added and 30 percent without it. However, additional work with both an amine and an ester of 2,4-D on this species demonstrated that the thiocyanate-enhanced effect could not be obtained consistently. Since the spray was applied at 20 gal./A, the thiocyanate concentration was, of course, exceedingly low in comparison with the other experiments described. It does seem obvious that considerably more knowledge is needed before inorganic salts can be recommended for spray formulations, particularly with herbicidal control of woody plants.

c) *pH*

There are at least two distinct systems within the leaf where hydrogen ion concentration may exert an influence on overall foliar absorption: the cuticle itself and the plasmatic membranes of underlying cells. Influence of pH on cuticular permeability may of course be examined in isolated cuticular membranes, whereas studies with isolated cells or tissue slices largely measure effect on plasmatic membranes. Many experimental techniques do not clearly differentiate these two systems, which may respond somewhat differently to changes in pH. When a pH-induced absorptive anomaly is noted following application to an intact leaf, it may be difficult to determine to what extent each of the systems is involved. The marked effect which pH has on solute absorption, both at the cellular level and in overall foliar penetration has been recognized for many years. It is generally considered that weak acids penetrate best at low pH values, where the molecules are largely undissociated. In this state they more readily penetrate the lipoidal phases of the cuticle and leaf cells. As pointed out by SIMON and BEEVERS (1952), the ionized components of biologically active molecules are often as active as the parent molecule. Thus the direct effect of pH on degree of dissociation of the active ingredient may be the primary cause for the effect of pH on activity, but responsible principally by virtue of its indirect effect on penetration. BRIAN (1964) determined changes in permeability of cellular membranes of *Nitella* from the electrical resistance of the membranes as measured by an internal micro-electrode. Changes in resistance proved to be a function of both type and concentration of the growth regulator present and pH of the bathing solution. At a pH of 3.8, where phenoxyacetic acids are largely undissociated, membrane resistance was variously increased by as much as 350 percent of the controls at concentrations of $0.6 \times 10^{-3} M$ and over. IAA was active in this manner only in the presence of Ca^{++}. In marked contrast, no significant change in resistance was observed at pH 6.0, where all regulators would be essentially in the dissociated form.

Not all solutes penetrate according to a simple inverse correlation with pH. Some of the anomalous responses to pH observed in the case of certain growth regulators and herbicides, as well as possible explanations for such responses, have been discussed by CRAFTS (1961), CURRIER and DYBING (1959), HULL (1964 a), VAN OVERBEEK (1956), SARGENT (1965), and others; they will not be further considered here. The interrelationship of pH to sorption of acid and basic dyes and ^{14}C-labeled 2,4-D and urea by isolated leaf cuticle has been thoroughly examined by ORGELL (1957), and pH effect on foliar absorption of mineral nutrients has been described by JYUNG and WITTWER (1965). An interesting point brought up by several of the above investigators is the fact that although penetration of the active ingredient is generally greater at low pH values, it is sometimes possible to enhance penetration at higher pH's upon addition of certain inorganic ions. Penetration may thus approach or perhaps equal that achieved at lower pH values. Attempts to enhance penetration additionally at low pH's in this manner have proven more difficult, the interrelationships becoming quite complex.

A marked effect of pH on penetration and translocation of 2,4-D was demonstrated by CRAFTS (1961) with the bean bend test. Maximum epinasty of the stem two hours following application to a primary leaf occurred at pH 2. By the sixth hour little difference in epinasty existed between pH values of two and six, although higher pH levels caused relatively little epinasty. The studies of GREENHAM (1968) on *Chondrilla juncea* show a generally similar pH effect. He measured foliar absorption and transport of ^{14}C-2,4-D by determining ^{14}C activity in the roots ten to 12 days after application. Significantly more activity was found at pH 3.5 than at 4.5, 6.5, or 8.5, although at 8.5 the inclusion of triethanolamine with the 2,4-D significantly increased its absorption and transport. Foliar absorption of dalapon by maize leaves was likewise found by FOY (1963) to be greatest from an aqueous solution of low pH, where the molecules were largely undissociated. However, acute toxicity at very low pH's created an opposing trend, so that most effective absorption was obtained at intermediate pH levels of about six. CRAFTS (1961) notes that pH apparently has little effect on penetration of MH. Variations in pH responsiveness of this nature doubtless exist within other classes of pesticides. A factor which may well influence the magnitude of pH-enhancement obtainable is the inherent phytotoxicity of the active ingredient and the concentration at which it is used. In spite of the numerous demonstrations of pH-enhanced absorption in laboratory or greenhouse experiments, relatively little benefit of pH control has been shown under field conditions. The somewhat heavier foliage and thicker cuticles common to field specimens may minimize the effect of pH to some extent. Also, little is known as to the actual depth within the foliage to which the pH of a spray solution may exert its influence. Clearly, much more work is needed before recommendations for practical application can be made.

d) *Droplet size*

For a subject of considerable importance, the effect of droplet size *per se* on biological activity has not received a great deal of study by pesticide investigators. Foliar retention is of course closely related to the nature of the plant surface and the physical characteristics of the spray itself. These relationships have been considered in previous sections of this review and have received comprehensive examination by EBELING (1963), HARTLEY (1966) and HOLLY (1964). Retention is additionally a function of droplet size. Of the research that has been accomplished in this area, much of it has been carried out with the objective of controlling droplet size in an effort to minimize drift. Several investigators have now demonstrated, however, that biological activity resulting from a specific spray formulation may be considerably altered simply by the adjustment of droplet size alone. In the present discussion both the physical aspects and possible biological influences of droplet size will be briefly considered. Information pertaining to the droplet size of emanations from different types of pesticide-application equipment, as well as the relationship between mean droplet diameter and the percentage of area covered by a gallon of spray solution (droplet contact angle 90°), has been made available by EBELING (1963).

BLACKMAN et al. (1958) examined in considerable detail the interrelations between spray retention and selective phytotoxicity. By using spray jets of varying sizes they were able to adjust both droplet size and spray output. Approximately equal volumes of output could be achieved by means of repeated applications in the case of the finer sprays. In Helianthus annuus, which has broad horizontal leaves, droplet size did not greatly influence retention at outputs up to about 75 ml./m². At higher outputs, retention was less with the larger droplet sizes. Hordeum vulgare, with a heavier cuticle and upright leaves, showed increased retention with smaller droplets throughout an output range of about 25 to 150 ml./m². Interrelationships of surface tension of the spray solution and droplet size with retention were studied by BRUNSKILL (1956). Water, with a surface tension of 71 to 72 dynes/cm.², dropped rapidly in its retentive properties on pea leaflets at droplet diameters above 100 μ and was almost completely reflected at diameters of 250 to 350 μ. Reduction in surface tension of such larger-sized droplets to about 45 dynes/cm.² or less resulted in about a ten-fold increase in retention. The angle of incidence at which such large drops impinged on the leaf surface, in relation to their retention, was also critically related to surface tension. At 42.5 dynes/cm.², the percentage of drops retained fell from 95 to 59 between angles of zero and 60 degrees, whereas at the slightly higher tension of 46.5 the percentage retention over the same angular range fell from 38 to one. The energetics involved in the process of droplet impaction onto artificial surfaces and the division of this process into stages of initial spread and subsequent retraction have been studied by FORD and FURMIDGE (1967 a). Degree of spreading immediately after impact was related to impact energy, as well as being a function of contact angle and surface tension. Subsequent retraction was governed both by extent of initial spread and by hysteresis of the contact angle.

Possible advantages of thickening agents in the spray formulation, such as hydroxyethyl cellulose (HEC), have already been discussed. LEHMAN et al. (1968) found that inclusion of HEC at concentrations between 0.5 and 1.0 percent significantly reduced the percentage of spray volume ejected as droplets less than 200 μ in diameter. Nozzle orientation on the boom of the fixed-wing aircraft also influenced the proportion of droplets of this size. The use of water-in-oil or "inverted" emulsions to minimize drift has also been mentioned. It was found by FORD and FURMIDGE (1967 b) that an increase in the viscosity of such emulsions could lead to either an increase or decrease in droplet size. Factors influencing the limiting value of the viscosity of invert emulsions, measured under conditions of very high shear, were ratio of the two phases and degree to which they were mixed, method of mixing, and the nature and concentration of both the toxicant and emulsifier used. Within limits, varying the phase ratio provided a convenient means of adjusting viscosity; it could also be changed by variations in the emulsifiers and toxicants. Since additional mixing occurs during emission from the spray nozzle, the selection of a specific mixing method as a way of adjusting viscosity was found to be somewhat impractical. Water-in-oil emulsions have proven particularly useful in minimizing drift of herbicide and certain other pesticide sprays. Biological activity resulting from the toxicants carried in such sprays is for the most part not greatly different from that achieved with "normal" oil-in-water emulsions, although some exceptions apparently do exist. Studying the effect of both normal and invert emulsions (1:7 oil-water ratio) as carriers for the aerial application of 2,4,5-T, LEHMAN et al. (1964) found that the spray distribution across the width of the swath was more uniform with the normal emulsion. Resultant defoliation of Prosopis juliflora and Quercus stellata which were being sprayed was likewise more homogeneous with the normal emulsion.

As to how droplet size and retention may influence subsequent biological activity of certain herbicides, probably the most comprehensive study to date is that of BENGTSSON (1961). Leaf wettability was evaluated by measuring droplet contact angles, which generally ranged from 60° to 160° on the numerous species investigated. Leaves of hard-to-wet plants (contact angle about 160°) showed relative retention ratios of 3:2:1 when sprayed with small (92 μ), medium (205 μ), and large (560 μ) droplets, respectively. With contact angles of about 70°, droplet size had little effect on retention. Also, it had little effect on retention at spray volumes of 160 to 600 l./ha., or at liquid surface tensions within the range of 35 to 70 dynes/cm.[2] In some plant species, the herbicidal effect of MCPA was considerably greater with small than it was with large droplet size; whereas in other species, droplet size had almost no effect on toxicity. These differences were primarily associated with wettability, the effect of droplet size being greatest at low volume applications and in species whose leaves were hardest to wet.

With diquat and paraquat, a somewhat different but still very significant effect of droplet size was observed by DOUGLAS (1968). Droplets ranging in diameter from 250 to 1,000 μ containing various concentrations of these herbicides were examined for toxicity toward Vicia faba. An increase in droplet size above 250 μ increased efficiency, as expressed in total lesion area/μg. of chemical applied. Optimum effect was reached between 400 and 500 μ, whereas activity fell off at still greater diameters of up to 1,000 μ. The specific influences of droplet size and spacing, spray volume, and herbicide concentration have been studied by BEHRENS (1957) as they relate to effectiveness of 2,4,5-T toward cotton and Prosopis juliflora. In the case of a water carrier deposited at a rate of 11 droplets/cm.[2] and 2,4,5-T rate held constant at

0.30 lb./A, there was a positive correlation between percentage of *Prosopis* stem tissue killed and droplet diameters ranging from 200 to 800 μ. However, droplet spacing proved to be of even greater importance than size. With a deposition rate of 11 droplets/cm.^{2}, an average spacing of 3.1 mm. proved to be the maximum distance compatible with optimum herbicidal effectiveness.

VIII. Conclusion

The pesticide applicator now has at his disposal an impressive arsenal of chemicals, each designed to accomplish a highly specific objective. A few years ago, the simple selection of the most effective active ingredient generally fulfilled the basic requirement for the spray job. From some of the above-described reports it now becomes evident that a further increase in degree of activity and selectivity can be expected if careful attention is paid to type and concentration of oil, surfactant, or other adjuvant, droplet size, time and method of application, etc. In the case of herbicides such selectivity may involve adjustments to the spray formulation which would enhance retention and subsequent toxicity to the weed species involved, while simultaneously promoting maximum runoff on the associated crop plant. Improved activity and selectivity of other types of pesticides should likewise be possible. For example, a fungicide or systemic insecticide formulation must be designed in such a way as to prevent serious injury to the host plant or associated plants, while still allowing for maximum activity against the target organism.

Another bonus becomes possible when pesticide activity is enhanced by the addition of specific adjuvants or the manipulation of droplet size and certain other of the peripheral factors which have been discussed. Some experiments have now clearly demonstrated that the magnitude of enhancement is sufficient in many cases to still achieve optimal degrees of biological activity with substantially reduced rates of active ingredient, thus resulting not only in economic savings but also a reduction of residues in plant and soil. Of course mammalian toxicities and biodegradability of the adjuvants must be considered in relation to similar properties of the pesticide itself, but rarely will they need to be of much concern.

Research with pesticides has for the most part centered on the various activities and functions of the toxicant itself, with relatively little effort having been directed toward the numerous peripheral factors which have also been shown to influence its activity and selectivity. Clearly, considerably more research is needed in this area, along with studies on the nature of the leaf surface and its interior ultrastructure. The coordination of information gained from these studies should increase our knowledge of foliar penetration in a way that should be of more than purely academic interest.

Acknowledgments

The author is indebted to the following for providing unpublished information or furnishing manuscripts prior to their publication: DRS. D. E. BAYER, C. A. BEASLEY, C. E. CRISP, D. A. FISHER, C. L. FOY, W. FRANKE, R. F. NORRIS, L. W. SMITH, and S. H. WITTWER. Grateful acknowledgment is also due DR. and MRS. F. A. GUNTHER, DRS. G. M. LOPER and H. L. MORTON, and MR. S. J. SHELLHORN for critical review of the manuscript and helpful suggestions. Special thanks are due MRS. BONNIE MCCLELLAN for editorial assistance and typing of the manuscript.

Table I. *Common and chemical names of pesticides and other chemicals mentioned in text*

Common name [a]	Chemical name
actinomycin D	$C_{62}H_{86}N_{12}O_{16}$
Alar	succinic acid 2,2-dimethylhydrazide
ametryne	2-(ethylamino)-4-(isopropylamino)-6-(methylthio)-*s*-triazine
amitrole	3-amino-*s*-triazole
amitrole-T	a mixture of amitrole and ammonium thiocyanate
ATP	adenosine triphosphate
Atplus® 300	polyoxyethylene sorbitan fatty acid ester
atratone	2-(ethylamino)-4-(isopropylamino)-6-methoxy-*s*-triazine
atrazine	2-chloro-4-(ethylamino)-6-(isopropylamino)-*s*-triazine
Azodrin®	dimethyl phosphate of 3-hydroxy-*N*-methyl *cis*-crotonamide
B-9	see Alar
B-995	see Alar
BA	N^6-benzyladenine
6-BAP	6-benzylaminopurine
BLO®	γ-butyrolactone
Bordeaux mix	$CuSO_4 \cdot 3Cu(OH)_2 \cdot H_2O$
bromoxynil	3,5-dibromo-4-hydroxybenzonitrile
Burgundy mix	$CuCO_3 \cdot Cu(OH)_2$ and/or $CuSO_4$
$(BuS)_3PO$	S,S,S-tributylphosphorotrithioate
Carbowax® 1500	polyethylene glycol
CCC	(2-chloroethyl)trimethylammonium chloride
chloramphenicol	D(-)-threo-2,2-dichloro-*N*-[β-hydroxy-α-(hydroxymethyl)-*p*-nitrophenethyl]acetamide
3-CP	3-chlorophenoxy-α-propionic acid
2-CPA	2-chlorophenoxyacetic acid
CTAB	trimethylammonium bromide
2,4-D	(2,4-dichlorophenoxy)acetic acid
2,4-DB	4-(2,4-dichlorophenoxy)butyric acid
dalapon	2,2-dichloropropionic acid
demeton	O,O-diethyl O-[2-(ethylthio)ethyl]phosphorothionate (isomeric mix)
dicamba	3,6-dichloro-*o*-anisic acid
dichlorprop	2-(2,4-dichlorophenoxy)propionic acid
dinocap	crotonic acid 2-(1-methylheptyl)-4,6-dinitrophenyl ester
dinoseb	2-*sec*-butyl-4,6-dinitrophenol
diquat	6,7-dihydrodipyrido[1,2-*a*:2',1'-*c*]pyrazinediium salts

Table I. (continued)

Common name	Chemical name
diuron	3-(3,4-dichlorophenyl)-1,1-dimethylurea
DMSO	dimethyl sulfoxide
DNOC	4,6-dinitro-o-cresol
DNP	2,4-dinitrophenol
dodine	n-dodecylguanidine acetate
DSMA	disodium methanearsonate
Dursban®	phosphorothioic acid O,O-diethyl-O-3,5,6-trichloro-2-pyridyl ester
endothall	7-oxabicyclo[2.2.1]heptane-2,3-dicarboxylic acid
EPTC	S-ethyl dipropylthiocarbamate
flurenol	n-butyl-9-hydroxyfluorene-(9)-carboxylate
GA	gibberellic acid
hexachlorophene	2,2'-methylenebis[3,4,6-trichlorophenol]
IAA	indole-3-acetic acid
ioxynil	4-hydroxy-3,5-diiodobenzonitrile
kinetin	N^6-furfuryladenine
linuron	3-(3,4-dichlorophenyl)-1-methoxy-1-methylurea
Lovo®	an amine stearate
MCPA	[(4-chloro-o-tolyl)oxy]acetic acid
MCPB	4-[(4-chloro-o-tolyl)oxy]butyric acid
MH	1,2-dihydro-3,6-pyridazinedione
monuron	3-(p-chlorophenyl)-1,1-dimethylurea
M-Pyrol®	N-methyl-2-pyrrolidone
MSMA	monosodium methanearsonate
NAA	naphthaleneacetic acid
NAAm	naphthaleneacetamide
naptalam	N-1-naphthylphthalamic acid
NMSP	a-1(naphthylmethylthio)propionic acid
paraquat	1,1'-dimethyl-4,4'-bipyridinium salts
parathion	O,O-diethyl O-p-nitrophenyl phosphorothioate
Phosphon	2,4-dichlorobenzyltributylphosphonium chloride
picloram	4-amino-3,5,6-trichloropicolinic acid
POA	phenoxyacetic acid
prometone	2,4-bis(isopropylamino)-6-methoxy-s-triazine
prometryne	2,4-bis(isopropylamino)-6-(methylthio)-s-triazine
propazine	2-chloro-4,6-bis(isopropylamino)-s-triazine
Rogor®	O,O-dimethyl S-methylcarbamoylmethyl phosphorodithioate
silvex	2-(2,4,5-trichlorophenoxy)propionic acid
simetryne	2,4-bis(ethylamino)-6-(methylthio)-s-triazine
streptomycin	$C_{21}H_{39}N_7O_{12}$
Synthalin®	decamethylenediguanidine dihydrochloride
2,4,5-T	(2,4,5-trichlorophenoxy)acetic acid
2,3,6-TBA	2,3,6-trichlorobenzoic acid
TCA	trichloroacetic acid
TIBA	2,3,5-triiodobenzoic acid
Triton® X-100	isooctylphenylpolyethoxy ethanol
Tween® 20	polyoxyethylene(20)sorbitan monolaurate
Tween® 40	polyoxyethylene(20)sorbitan monopalmitate
Tween® 80	polyoxyethylene(20)sorbitan monooleate
X-77, Multi-Film®	alkylarylpolyoxyethylene glycols, free fatty acids and isopropanol

ᵃ Trade names are capitalized and with ® symbol.

Table II. *Glossary of botanical terms*

Term	Definition
Abscission	The separation of leaves, flowers, and fruit from plants by the development and subsequent disorganization of the separation layer
Acropetal	From the base toward the apex
Apoplast	The portion of cellular tissue exterior to the living protoplasts, through which a substance may move en masse or by diffusion. It includes essentially the cuticle, the cell walls, and the xylem vessels in which the transpiration stream moves
Appressorium	The flattened, thickened, or tuftlike tip of a hyphal branch by which certain parasitic fungi are attached to their hosts
Anticlinal	Situated at right angles to the surface
Astomatous	Without stomata
Basipetal	From the apex toward the base
Border parenchyma	See bundle sheath
Bundle sheath	A layer of compactly arranged parenchyma cells surrounding the vascular bundles in leaves
Callose	A carbohydrate component of cell walls that is readily stained by aniline blue. It is often found on sieve plates where it may cause blockage.
Cladophyll	A branch which assumes the form of and closely resembles an ordinary foliage leaf, and is borne in the axil thereof
Coleoptile	The first leaf of a monocotyledonous plant which forms a protective sheath about the plumule
Companion cell	An elongated parenchyma cell within the phloem which lies next to the sieve tube, and with which it is apparently physiologically associated
Conidia	Asexual spores produced by abstriction, budding, or septation from the tip of a conidiphore
Corolla	Flower petals, collectively; usually the conspicuous colored flower whorl
Cortical cells	Cells making up the tissue of a stem or root which are bounded externally by the epidermis and internally by the pericycle
Cotyledon	The leaf or leaves of the plant embryo
Cultivar	A plant variety or strain that has persisted under cultivation
Cuticle	The layer on the outer surface of epidermal cells consisting predominantly of wax, cutin, and pectin components
Cystolith	A calcium carbonate concretion, commonly stalked and arising from the cellulose wall of certain cells of higher plants, particularly epidermal cells
Dendritic	Branched, resembling a tree in form
Dicot	Short for dicotyledon; a plant whose embryo has two cotyledons
Ectodesmata	Microscopic tube-shaped structures found in the epidermal cell walls of numerous plants, resembling plasmodesmata, but apparently nonplasmatic in nature

Table II. (continued)

Term	Definition
Endoplasmic reticulum	A network of membrane-bound cavities, or cisternae, within the interior of the cytoplasm
Epicuticular	Refers to the outermost waxy layer of the cuticle, including the surface wax
Epidermis	The outermost layer of cells of plant organs
Epinasty	A nastic movement by which a plant part is bent outward and often downward
Fiber cell	An elongate tapering cell having a thick wall and little or no protoplasm, found commonly in the xylem and phloem of the vascular system
Germ-tube	The slender tubular outgrowth first produced by most spores in germination
Glabrous	Without pubescence; hairless
Ground parenchyma	Fundamental or ground tissue, composed of cells having thin walls and a polyhedral shape, and concerned with vegetative activities
Guard cell	One of the two crescent-shaped epidermal cells united at their ends whose changes in turgidity determine opening and closing of a stoma
Guttation	The exudation of moisture from an uninjured surface of a plant, as from a hydathode
Gymnosperm	A plant of the class Gymnospermae, comprising seed plants that produce naked seeds not enclosed in an ovary, such as the genus *Pinus*
Haustorium	Projections of hyphae which act as penetrating and food-absorbing organs
Heteroblastic	Having an indirect embryonic development or having young and adult forms different, as in the case of leaves
Hyaloplasm	The clear apparently homogeneous ground substance of cytoplasm that is essentially the continuous phase of a multiple-phase colloidal system
Hydathode	An epidermal opening resembling a stoma which serves for the guttation or exudation of water
Hydrophyte	A vascular plant growing wholly or partially in water
Hypocotyl	The portion of the axis of a plant embryo or seedling below the cotyledons
Hypodermal cells	A layer of cells directly under the epidermal cells
Intercostal	Situated between the veins of a leaf
Lamina	The blade or expanded part of a foliage leaf
Meristem	Undifferentiated tissue, the cells of which are capable of active division
Mesocotyl	The elongated portion of the axis between the cotyledon and the coleoptile of a grass seedling
Mesophyll	Parenchyma tissue between the epidermal layers of a foliage leaf
Mitochondria	Granular or filamentous bodies within the protoplast having genetic continuity. They are apparently capable of dividing and contain some of the principal oxidative enzymes
Monocot	Short for monocotyledon; a plant whose embryo has one cotyledon

Table II. (continued)

Term	Definition
Organelle	The endoplasmic reticulum, dictyosomes, and other specialized components of cellular protoplasm which perform specific functions
Palisade cell	A columnar or cylindrical parenchyma cell, rich in chloroplasts, and generally found immediately beneath the upper epidermis of foliage leaves
Papillae	Small nipple-shaped projections
Parenchyma	Thin-walled cells that serve for photosynthesis and/or storage, and that make up much of the substance of leaves, stems, roots and fruits
Peltate	Referring to a structure which has a stem or other support attached to its lower surface rather than to its base or margin
Perianth	Petals and sepals, collectively
Periclinal	Parallel with the surface
Periderm	A protective layer of secondary tissue that develops first in the epidermis or subepidermal layers of various plant organs. At full development it consists of an initiating layer, an inner parenchymatous layer, and an outer cork layer
Petiole	The stalk of a leaf
Phloem	A tissue in the vascular system of higher plants consisting mainly of sieve tubes and companion cells, and often containing fiber and parenchyma cells. It functions in translocation, support, and storage
Phloem parenchyma	A nonspecialized, vertically arranged series of parenchyma cells within the phloem
Phytochrome	A light-receptive pigment in plants responsible for numerous physiological responses
Plasmalemma	See plasma membrane
Plasma membrane	A semipermeable membrane between the protoplasm and the cell wall
Plasmodesmata	Tube-like structures within cell walls containing intercellular protoplasmic connections
Plastid	A small body of specialized protoplasm lying within the cellular cytoplasm
Protoplast	The living content of a cell, including the cytoplasm, nucleus, and plasma membrane
Pteridophyte	A member of the Pteridophyta, a division of the plant kingdom including ferns and their allies
Pubescent	Covered with fine short hairs
Pulvinus	A cushionlike enlargement of the base of a petiole or petiolule, consisting of a mass of thin-walled cells surrounding a vascular strand
Reticulum	A netlike structure
Rhizome	An elongate stem or branch, usually horizontal and underground, and producing shoots above and roots below
Ribosome	A submicroscopic granule occurring free in the cytoplasm or associated with endoplasmic reticulum, and apparently a macromolecule of ribonucleoprotein that participates in protein synthesis

Table II. (continued)

Term	Definition
Russeting	A brownish roughened area on fruit skin caused by frost, insect or fungus injury, or spraying
Sieve element	A transport element of phloem, whether taking the form of a sieve cell or a sieve-tube element
Sieve plate	A wall or portion of a wall between sieve-tube elements, containing one or more sieve areas
Solanaceous	Referring to the family Solanaceae
Spicule	A small spikelike prominence
Spongy parenchyma	Loosely and irregularly arranged parenchyma, often with large intercellular spaces, and found toward the lower leaf surface
Stomata	Minute openings in the foliar epidermis through which gaseous interchange between the atmosphere and the intercellular spaces occurs
Stomatous	Having stomata
Symplast	The living protoplasm, including the protoplasts and their intercellular connections via plasmodesmata; also, the interconnecting protoplasm of the sieve elements
Tonoplast	The semipermeable plasmatic membrane surrounding the vacuole of plant cells
Trichome	An epidermal hair
Umbelliferous	Referring to the family Umbellifereae
Vacuole	A cavity in the cytoplasm containing cell sap and surrounded by a semipermeable membrane, the tonoplast
Vascular bundle	A unit strand of the vascular system of a higher plant, generally consisting of vessels and sieve elements commonly in association with elongated parenchyma cells and fibers that in some cases surround the strand as a sheath
Vesicle	A small and more or less circular elevation of the cuticle containing a clear watery fluid
Xeromorphic	Relating to climatic conditions favorable for the development of xerophytic vegetation
Xerophyte	A plant structurally adapted for life and growth with a limited water supply
Xylem	A tissue in the vascular system of higher plants consisting of vessels, tracheids, or both, usually together with fiber and parenchyma cells, and which functions in conduction, support, and storage

Summary

It has become increasingly evident that effective use of foliar-applied pesticides requires a knowledge of the rather intricate relationship existing between surface and interior leaf structure and chemical and physical characteristics of the formulation applied. For this reason, both established and theoretical concepts of ultrastructure have been examined as they relate to foliar penetration. The cuticle itself is difficult to study ultrastructurally due to the problem of sharply differentiating the various cuticular components with electron-dense stains, and because of the tendency for the waxy portions to volatilize under the high vacuum and temperatures imposed by the electron beam. Recent findings relating to wax biosynthesis are discussed; for example, the apparent pathway for paraffin biosynthesis which involves elongation of a common fatty acid to a very long chain of appropriate length, followed by decarboxylation.

Preferred routes of penetration, *i.e.*, stomatal or cuticular, are shown to be interrelated with plant species and the nature of the penetrant. Entry via the trichomes, which may be considered a type of cuticular penetration, becomes of predominant importance in certain species and at specific stages of foliar development. As suggested by fluorochrome penetration, there is evidence that optimum requirements for surface tension and other characteristics of the solution are somewhat different for penetration via the trichomes than they are for the stomata.

Recent ultrastructural evidence relating to the controversial subject of cuticular pores has contributed to our knowledge of these structures. First-stage replication obtained by vacuum infiltration of low-viscosity plastics has indeed demonstrated the presence of pores in the species examined. Their function as pathways for the extrusion of epicuticular wax is suggested by the association of such surface wax crystals with individual or clusters of underlying pores. The role which they may play, if any, in cuticular transpiration or as portals for foliar penetration is less certain.

Additional evidence has been obtained of the important part played by ectodesmata in foliar absorption and excretion. The presence of ectodesmata, however, existing as actual structures prior to fixation with mercuric chloride has been questioned. Such doubt is based on the ephemeral nature of the structures, the difficulty or impossibility of demonstrating them in many species, and the fact that the corrosive sublimate necessary for such demonstration may cause varying degrees of dissolution of the cellulose in the epidermal cell walls. The possibility apparently exists that the structures visible by light and electron microscopy are principally the result of such dissolution. However, this hardly seems possible in view of the fact that the sublimate preferentially penetrates specific loci within the wall, thus suggesting an anomalous structure of some type. Ectodesmata have also been demonstrated by a silver iodide method. Their role in foliar penetration is also considered in relation to new evidence about cuticular regions which show variations in ion binding capacity and permeability.

Of the various plant and environmental factors influencing penetration, high temperature and atmospheric humidity are especially important factors favoring penetration. The effects of light and soil moisture stress are less clear. Although stress severely inhibits translocation, its influence on absorption is complex and still controversial.

Numerous types of adjuvants have been tested for use with pesticide spray formulations. Many give erratic results or only borderline enhancement of biological activity. Others are highly effective in increasing the activity of some pesticides under specific conditions. Surfactants, oils, deposit builders, and penetrants such as dimethyl sulfoxide, have proven distinctly beneficial when properly selected, formulated, and applied. Regulation of these peripheral factors often permits equivalent degrees of biological activity to be achieved with reduced quantities of the active toxicant, thereby minimizing environmental contamination. Markedly enhanced degrees of selectivity have likewise been demonstrated.

In spite of greatly expanded research on pesticides in recent years, serious gaps exist in several areas. Considerable work has been concentrated on the active ingredient itself, whereas problems relating to certain of the peripheral factors which have been described, and which influence the activity of this ingredient, have not received the attention they deserve. For example, the surprising extent to which pesticide activity and selectivity can be altered simply by the adjustment of spray droplet size needs additional investigation. Thus, in the case of herbicides, activity resulting from small spray droplets may be considerably greater on some species than that resulting from larger droplets; with other species droplet size has no significant effect. Regulation of pesticide sprays in this manner, in areas where spray drift is not critical, should help achieve the desired goals of increased selectivity and minimized residue to plants and soil.

Résumé*

Rôle de la structure des feuilles dans l'absorption de pesticides et autres composés

Il est devenu de plus en plus évident que l'emploi efficace des pesticides applicables à la surface des feuilles exige une connaissance des rapports assez complexes qui existent entre la structure externe et interne de la feuille et les caractéristiques chimiques et physiques de la formulation appliquée. C'est pourquoi les concepts théoriques, aussi bien que les concepts d'ultrastructure déjà établis, ont été examinés en tant qu'ils se rapportent à la pénétration foliaire. La cuticule elle-même est difficile à étudier de façon ultrastructurale à cause de la difficulté qui se présente à différencier nettement les constituants cuticulaires par le moyen de taches à grande densité d'électrons, et à cause de la tendance qu'ont les éléments cireux à se volatiliser en présence du haut degré de vide et de température créé par le faisceau électronique. Les découvertes récentes concernant la biosynthèse de la cire sont traitées, par exemple, la voie apparente de la biosynthèse de la paraffine, où il s'agit de l'allongement d'un acide gras commun pour former une chaîne très étendue de longueur convenable, suivi de décarboxylation.

---

* Traduit par M. Rodack.

Il est démontré que les routes de pénétration préférées, c'est-à-dire, par voie des stomates ou de la cuticule, ont des rapports étroits avec l'espèce végétale et la nature du pénétrant. L'entrée par voie des trichomes, qu'on peut considérer comme une sorte de pénétration cuticulaire, devient particulièrement importante chez certaines espèces et dans certaines phases spécifiques du développement des feuilles. Comme l'indiquerait la pénétration par le fluorochrome, il y a des témoignages que les conditions requises optimums pour la tension superficielle et autres caractéristiques de la solution, sont assez différentes dans le cas de la pénétration par voie des trichomes qu'elles ne le sont pour les stomates.

Des témoignages ultrastructuraux récents, se rapportant au sujet polémique des pores cuticulaires, ont contribué à notre connaissance de ces structures. Une réplique primaire, obtenue par infiltration dans le vide de plastiques à basse viscosité, a en effet démontré la présence de pores dans les espèces étudiées. Leur fonction comme voies d'expulsion pour la cire épicuticulaire est suggérée par le fait que de pareils cristaux de cire superficielle sont associés aux pores sous-jacents isolés ou en groupe. Leur rôle, s'ils en ont un, soit dans la transpiration cuticulaire, soit comme portes de pénétration foliaire, est moins certain.

Des témoignages supplémentaires du rôle important joué par les ectodesmes dans l'absorption et l'excrétion foliaires ont été obtenus. La présence des ectodesmes, pourtant, existant comme structures positives avant leur fixation par le chlorure de mercure, a été mise en question dans une certaine mesure. Ce doute est basé sur la nature éphémère des structures, le fait qu'il est difficile, sinon impossible, de les démontrer chez beaucoup d'espèces, et le fait aussi que le sublimé corrosif nécessaire pour une telle démonstration peut causer, à un degré plus ou moins grand, la dissolution du cellulose dans les parois des cellules épidermiques. La possibilité semblerait exister que les structures visibles par moyen de la microscopie traditionnelle et électronique sont principalement les résultats de cette dissolution. Pourtant, ceci paraît à peine possible, étant donné que le sublimé pénètre de préférence des points spécifiques à l'intérieur de la paroi, indiquant ainsi une structure irrégulière de quelque sorte. Les ectodesmes ont aussi été démontrés par une méthode employant l'iodure d'argent. Leur rôle dans la pénétration foliaire est considéré aussi par rapport à de nouveaux témoignages concernant les régions cuticulaires qui montrent des variations dans leur perméabilité et leur capacité de capter les ions.

Parmi les différents facteurs propres aux plantes ou au milieu qui influencent la pénétration, les températures élevées et l'humidité atmosphérique sont particulièrement favorables. Les effets de la lumière et de la charge produite par l'humidité du sol sont moins nets. Bien que cette charge inhibe la translocation, son influence sur l'absorption est complexe et encore sujet à discussion.

De nombreux genres d'adjuvants ont été vérifiés en tant qu'éléments à employer avec les formulations de pesticides pulvérulants. Beaucoup donnent des résultats erratiques ou ne contribuent que d'une façon limitée à l'activité biologique. D'autres augmentent d'une façon très efficace l'activité de certains pesticides sous des conditions spécifiques. Les tensio-actifs, les huiles, les formateurs de dépôts, les pénétrants tels que le sulfoxyde de diméthyle, se sont montrés très salutaires quand ils sont convenablement choisis, formulés et appliqués. Le réglage de ces facteurs auxiliaires permet souvent d'atteindre un degré d'activité biologique également élevé avec une quantité moins grande du toxique actif, réduisant au minimum la contamination du milieu. Une sélectivité bien plus raffinée a aussi été démontrée.

Bien que les recherches dans le domaine des pesticides se soient multipliées depuis quelques années, il existe des lacunes importantes dans plusieurs départements. Beaucoup de travaux ont été consacrés à l'ingrédient actif lui-même, tandis que les problèmes relatifs à certains des facteurs auxiliaires qui ont été décrits, et qui influencent l'activité de cet ingrédient, n'ont pas reçu l'attention qu'ils méritent. Par exemple, la mesure étonnante dans laquelle on peut modifier l'activité et la sélectivité des pesticides par le simple réglage de la dimension des gouttelettes, exige des recherches supplémentaires. Ainsi, dans le cas des herbicides, l'activité produite par une pulvérisation à petites gouttelettes peut, chez certaines espèces, être beaucoup plus considérable que celle qui provient de gouttelettes plus grandes; chez d'autres espèces la dimension des gouttelettes n'a aucun effet significatif. Ce genre de réglage des pesticides pulvérulants dans les situations où l'entraînement par le vent n'est pas critique, devrait aider à atteindre les buts désirés—une plus grande sélectivité et la réduction au minimum du résidu dans les plantes comme dans le sol.

Zusammenfassung[*]

Blattstruktur in Beziehung zur Absorption von Pestiziden und anderen Verbindungen

Es ist immer klarer geworden, dass die wirksame Anwendung von Schädlingsbekämpfungsmitteln bei Pflazenblättern eine Kenntnis der ziemlich verwickelten Beziehungen zwischen äusseren und inneren Blattstrukturen, und den chemischen und physikalischen Merkmalen der angewandten Präparate erfordert. Deshalb wurden sowohl erforschte wie theoretische Konzepte der Ultrastruktur im Hinblick auf die Durchdringung des Blattwerkes überprüft. Es ist schwierig, die Kutikel selbst als Feinstruktur zu untersuchen, da infolge der elektrondichten Färbung das Differenzieren der verschiedenen Kutikularkom-

---

[*] Übersetzt von G. O. W. KREMP und E. M. KREMP,

ponenten problematisch wird, und da die wachsenthaltenden Teile die Tendenz haben, unter den hohen Vacuumbedingungen und unter den hohen Temperaturen, die durch die Elektronstrahlen hervorgerufen werden, sich Verflüchtigen. Neue Untersuchungsergebnisse werden in Verbindung mit der Wachs-Biosynthese diskutiert, so z.B. der vermutliche Entwicklungsgang der Paraffin–Biosynthese, der die Verlängerung einer gewöhnlichen Fettsäurekette zu einer sehr langen Kette von bestimmtem Ausmass mit sich bringt, und der dann die Dekarboxylierung folgt.

Es hat sich gezeigt, dass besondere Eindringungswege, z.B. durch Spaltöffnungen oder die Kutikula, von der Pflanzenart und der Natur des eindringenden Stoffes abhängig sind. Das Eindringen über die Trichome, welches als eine Art Kutikularinfiltration betrachtet werden kann, erlangt bei gewissen Arten und in bestimmten Stadien der Blattentwicklung entscheidende Bedeutung. Bezüglich der Fluorochrompenetration sind Hinweise vorhanden, dass die optimalen Bedingungen für Oberflächenenergie und andere Charakteristika der Lösung davon abhängen, ob das Eindringen über die Trichome oder durch die Spaltöffnungen geschieht.

Neue Untersuchungen von Ultrastrukturen im Hinblick auf das strittige Thema der kutikularen Poren hat zu unserem Verständnis dieser Poren beigetragen. Eine direkte Nachbildung der Blattoberfläche, die durch Vacuum-Infiltration von gering-viskosen Kunststoffen hergestellt wurde, hat tatsächlich in den untersuchten Arten die Anwesenheit von Poren gezeigt. Auf ihre Funktion als Durchgänge für die Ausscheidung von epicularem Wachs deutet die Verbindung von solchen Wachskristallen der Oberfläche mit einzelnen darunter liegenden Poren oder Gruppen von Poren hin. Ob sie jedoch eine Rolle in der kutikularen Transpiration spielen, und gegebenenfalls welche, und ob sie Eingänge für die Penetration der Blätter darstellen, ist nicht sicher.

Über die wichtige Rolle, die Ektodesmen bei den foliaren Absorbierungen und Ausscheidungen spielen, hat man neues Beweismaterial erhalten. Die Anwesenheit von Ektodesmen als tatsächliche Strukturen vor der Fixierung mit Quecksilberchlorid ist jedoch fraglich. Diesbezügliche Zweifel beruhen auf der kurzlebigen Natur solcher Strukturen, auf der Schwierigkeit oder Unmöglichkeit, sie in einer grösseren Anzahl von Arten zu zeigen, und auf der Tatsache, dass das für eine solche Darstellung notwendige ätzende Sublimat verschiedene Auflösungsgrade der Zellulose in den epidermalen Zellwänden hervorrufen kann. Offenbar besteht die Möglichkeit, dass die bei Licht- und Elektronmikroskopie sichtbaren Strukturen im wesentlichen das Resultat einer solchen Auflösung sind. Andererseits kann dies kaum möglich sein, in Anbetracht der Tatsache, dass das Sublimat gewöhnlich ganz bestimmte Stellen innerhalb der Wand durchdringt, und damit auf irgendeine anormale Struktur hinweist. Ektodesmen sind auch mit Hilfe einer Jod-Silbermethode dargestellt worden. Ihre Rolle bei der Durchdringung des Blattwerkes wird auch mit neuen Untersuchungen in Verbindung gebracht, die auf dem Gebiet der Kutikularforschung durchgeführt worden sind. Es hat sich gezeigt, dass verschiedene kutikulare Regionen Variationen in ionbindender Kapazität und verschiedene Permeabilität besitzen.

Von den verschiedenen Pflanzen- und Umweltsfaktoren, die Einfluss auf die Durchdringung haben, fördern hohe Temperaturen und atmosphärische Feuchtigkeit in besonderem Masse die Infiltration. Die Wirkung von Licht und von Bodenfeuchtigkeitsspannung ist weniger klar. Obwohl es bekannt ist, dass letztere Stoffverlagerungen innerhalb der Pflanze stark hemmt, ist ihr Einfluss auf Absorption kompliziert und noch immer strittig.

Zahlreiche Typen von Adjuvanten sind auf ihre Nutzanwendung in Pflanzen-Sprühschutzmitteln untersucht worden. Viele von ihnen liefern unberechenbare Resultate oder ergeben nur geringe Erhöhung biologischer Aktivität. Andere sind äusserst wirksam, indem, sie die Aktivität einiger Schädlingsbekämpfungsmittel unter besonderen Bedingungen erhöhen. Oberflächenaktive Substanzen, Öle, Viskose-Agenzien und eindringende Substanzen, wie z.B. Dimethylschwefeloxyd, haben sich eindeutig als vorteilhaft erwiesen, wenn sie sorgfältig ausgewählt, zusammengestellt und angewandt wurden. Die Regulierung dieser Nebenfaktoren erlaubt oftmals, equivalente Grade von biologischer Aktivität mit reduzierten Mengen des aktiven, toxischen Stoffes zu erreichen, und somit die Umweltsverunreinigung auf ein Minimum zu beschränken. Merkbar erhöhte Grade von Selektivität sind ebenfalls dargestellt worden.

Trotz der in den letzten Jahren erfolgten zahlreichen Untersuchungen, die sich mit Pflanzenschutzmitteln beschäftigt haben, bestehen doch schwerwiegende Lücken auf verschiedenen Gebieten. Beachtliche Arbeit ist in Bezug auf den aktiven Komponenten selbst geleistet worden; dahingegen ist Problemen in Verbindug mit einigen der Nebenfaktoren, die hier beschrieben worden sind, und die die Aktivität dieser Stoffe beeinflussen, nicht die ihnen gebührende Aufmerksamkeit erwiesen worden. So erfordert z.B. das erstaunliche Mass, in welchem die Wirksamkeit und die Selektivität der Pflanzenschutzmittel geändert werden kann, einfach indem man die Grösse der Sprühtropfen variiert, weitere Untersuchungen. Bei Kräutern zum Beispiel kann die Aktivität, die sich bei kleinen Sprühtropfen entfaltet, bei einigen Pflanzenarten beachtlich grösser sein, als die, die sich bei grösseren Tropfen entwickelt. Es gibt jedoch auch Pflanzenarten, bei denen die Tröpfchengrösse keine erhebliche Konsequenz hat. Eine derartige Regulierung von Sprühverfahren bei der Anwendung von Pflanzenschutzmitteln sollte dazu beitragen, das erwünschte Ziel einer erhöhten Selektivität und eines minimalen Rückstandes in Pflanzen und Boden zu erreichen; jedenfalls für Gebiete, in denen man nur mit unbedeutender Sprühregen-Drift zu rechnen braucht.

References

AHLGREN, G. E., and T. W. SUDIA: Absorption of P-32 by leaves of *Glycine max* of different ages. Bot. Gaz. **125**, 204 (1964).

—— —— Studies on the mechanism of the foliar absorption of phosphate. In: Proceedings of the symposium on isotopes in plant nutrition and physiology, p. 347. Sept. 5–9, 1966, Vienna. Internat. Atomic Energy Agency: Vienna, Austria (1967).

AHMAD, K. J.: Cuticular studies on Solanaceae. Can. J. Bot. **42**, 793 (1964).

AKHAVEIN, A. A., and D. L. LINSCOTT: The dipyridylium herbicides, paraquat and diquat. Residue Reviews **23**, 97 (1968).

ALBERSHEIM, P., and U. KILLIAS: Histochemical localization at the electron microscope level. Amer. J. Bot. **50**, 732 (1963).

AL'TERGOT, V. F., and V. E. KISELEV: O prirode sovmestnogo deistviya na pshenitsu i shirokolistnye sornyaki smesei rastvorov mineral'nykh solei i 2,4-D, vvodimykh cherez list. In: Fiziologicheskie osnovy priemov povyshenii ustoichivosti i produktivnosti rastenii v Sibiri. Sib. Otd. Akad. Nauk. SSSR: Novosibirsk, p. 149 (1963); through Ref. Zh. Biol. No. 3G125 (1965).

AMBLER, J. E.: Translocation of strontium from leaves of bean and corn plants. Radiat. Bot. **4**, 259 (1964).

——, and R. G. MENZEL: Retention of foliar applications of Sr[85] by several plant species as affected by temperature and relative humidity of the air. Radiat. Bot. **6**, 219 (1966).

AMBRONN, H.: Über das optische Verhalten der Cuticula und der verkorkten Membranen. Ber. Deut. Bot. Ges. **6**, 226 (1888).

AMELUNXEN, F., K. MORGENROTH, and T. PICKSAK: Untersuchungen an der Epidermis mit dem Stereoscan-Elektronenmikroskop. Z. Pflanzenphysiol. **57**, 79 (1967).

AMER, F. A., and W. T. WILLIAMS: Drought resistance in *Pelargonium zonale*. Ann. Bot. **22**, 369 (1958).

AMSDEN, R. C.: Reducing the evaporation of sprays. Agr. Aviat. **4**, 88 (1962).

——, and C. P. LEWINS: Assessment of wettability of leaves by dipping in crystal violet [weed control]. World Review Pest Control **5**, 187 (1966).

ANDERSON, D. B.: The distribution of cutin in the outer epidermal wall of *Clivia nobilis*. Ohio J. Sci. **34**, 9 (1934).

ARENS, T.: Radialstrukturen in den Stomata von *Ouratea spectabilis* (Mart.) Engl. Protoplasma **66**, 403 (1968).

ARISZ, W. H.: Influx and efflux of electrolytes. II. Leakage out of cells and tissues. Acta Bot. Neer. **13**, 1 (1964).

ARRAES, M. A. B., R. D. MACHADO, and V. A. NEPOMUCENO: Sobre a anatomia de folha da carnaubeira *Copernicia prunifera* (Miller) H. E. Moore. An. Acad. Brasil Cienc. **38**, 73 (1966).

ASHWORTH, R. DE B., and G. A. LLOYD: Laboratory and field tests for evaluating the efficiency of wetting agents used in agriculture. J. Sci. Food Agr. **12**, 234 (1961).

ÅSLANDER, A.: Sulphuric acid as a weed spray. J. Agr. Research **34**, 1065 (1927).

AUDUS, L. J.: The transport of growth regulators in plants. Agrochim. **11**, 309 (1967).

AYA, F. O., and S. K. RIES: Influence of oils on the toxicity of amitrole to quackgrass. Weed Sci. **16**, 288 (1968).

BADIEI, A. A., E. BASLER, and P. W. SANTELMANN: Aspects of movement of 2,4,5-T in blackjack oak. Weeds **14**, 302 (1966).

BAIN, J. M., and D. McC. McBEAN: The structure of the cuticular wax of prune plums and its influence as a water barrier. Austral. J. Biol. Sci. **20**, 895 (1967).

BAKER, E. A., R. F. BATT, and J. T. MARTIN: Studies on plant cuticle. VII. The nature and determination of cutin. Ann. Applied Biol. **53**, 59 (1964).
——, and J. T. MARTIN: Studies on plant cuticle. X. The cuticles of plants of related families. Ann. Applied Biol. **60**, 313 (1967).
BANCHER, E., J. HÖLZL, and J. KLIMA: Licht- und elektronenmikroskopische Beobachtungen an der Kutikula der Zwiebelschuppe von *Allium cepa*. Protoplasma **52**, 247 (1960).
—— ——, and R. SCHIFFAUER: Zum Mechanismus der stoffwechselbedingten Vakuolenfluorochromierung mit Uranin. Protoplasma **66**, 327 (1968).
BANDEEN, J. D., and W. C. VERSTRAETE: Effects of several oils and surfactants on the enhancement of atrazine for weed control in corn. Weed Soc. Amer. Abstr., p. 9 (1967).
BARRIER, G. E., and W. E. LOOMIS: Absorption and translocation of 2,4-dichlorophenoxyacetic acid and P^{32} by leaves. Plant Physiol. **32**, 225 (1957).
BASLER, E., G. W. TODD, and R. E. MEYER: Effects of moisture stress on absorption, translocation, and distribution of 2,4-dichlorophenoxyacetic acid in bean plants. Plant Physiol. **36**, 573 (1961).
BATT, R. F., and J. T. MARTIN: The cuticle of Cox's Orange Pippin apples from a manurial trial. J. Hort. Sci. **41**, 271 (1966).
BAUR, J. R., R. W. BOVEY, P. S. BAUR, and Z. EL. SEIFY: Effects of paraquat on the ultrastructure of mesquite mesophyll cells. In: Brush research in Texas, p. 68. Texas Agr. Expt. Sta., College Station, Texas (1968).
BAYER, D. E., and H. R. DREVER: The effects of dimethyl sulfoxide on absorption and translocation of dalapon and diuron. W. Weed Contr. Conf. Research Progress Rept., p. 106 (1964).
—— —— The effects of surfactants on efficiency of foliar-applied diuron. Weeds **13**, 222 (1965).
——, and S. YAMAGUCHI: Absorption and distribution of diuron-C^{14}. Weeds **13**, 232 (1965).
BEASLEY, C. A., and R. M. PHILPOT: Personal communication (1969).
BEHRENS, R.: Influence of various components on the effectiveness of 2,4,5-T sprays. Weeds **5**, 183 (1957).
—— The physical and chemical properties of surfactants and their effects on formulated herbicides. Weeds **12**, 255 (1964).
BENDA, G. T. A.: Vacuolar mass-flow in the floral hairs of *Stapelia*. Amer. J. Bot. **51**, 778 (1964).
BENGTSSON, A.: Droppstorlekens inflytande på ogräsmedlens verkan. Växtodling **17**, 1 (1961).
BENSON, A. A.: Plant membrane lipids. Ann. Rev. Plant Physiol. **15**, 1 (1964).
BISWAS, P. K., and D. D. HEMPHILL: Role of growth regulators in the uptake and metabolism of s-triazine herbicide by tea leaves. Nature **207**, 215 (1965).
——, and M. N. ROGERS: Effects of hygroscopic agents and pre-hydration of the leaves on absorption of gibberellic acid by column stock, *Matthiola incana*. Indian J. Plant Physiol. **6**, 34 (1963).
BLACK, F. S.: New concept for spray tank adjuvants. N.C. Weed Control Conf. Proc., p. 39 (1968).
BLACKLOW, W. M., and D. L. LINSCOTT: The fate of 2,4-D applied to Viking birdsfoot trefoil and a resistant intercross. Weed Sci. **16**, 516 (1968).
BLACKMAN, G. E., R. S. BRUCE, and K. HOLLY: Studies in the principles of phytotoxicity. V. Interrelationships between specific differences in spray retention and selective toxicity. J. Expt. Bot. **9**, 175 (1958).
BLUMBERG, A. A., and P. A. ILAGAN: Glucose transport through cellulose acetate membranes: The influence of dimethyl sulfoxide. J. Colloid Interface Sci. **27**, 681 (1968).
BOLHÀR-NORDENKAMPF, H.: Zur Physiologie der Fluorescein-Speicherung in Pflanzenzellen. Protoplasma **62**, 113 (1966).

BOLIS, L., V. CAPRARO, K. R. PORTER, and J. D. ROBERTSON: Symposium on bio-physics and physiology of biological transport. Proc. 1965 Meeting at Frascati. Protoplasma 63 (1–3), special ed. (1967).

——, and B. A. PETHICA: Membrane models and the formation of biological membranes. Proc. 1967 Meeting Internat. Conf. Biol. Membranes. Amsterdam: North Holland Publishing Co.; New York: Wiley (1968).

BOROUGHS, H., and C. LABARCA: El uso de humectantes en nutrición foliar. Turrialba 12, 204 (1962).

BOROVYAGIN, V. L.: Sovremennoe predstavlenie o strukture kletochnykh membran. Tsitol. 9, 505 (1967).

BOVEY, R. W., and F. R. MILLER: Phytotoxicity of paraquat on white and green hibiscus, sorghum and alpinia leaves. Weed Research 8, 128 (1968).

BOWEN, J. E.: Borate absorption in excised sugarcane leaves. Plant & Cell Physiol. 9, 467 (1968).

BRACKER, C. E.: Ultrastructure of the haustorial apparatus of Erysiphe graminis and its relationship to the epidermal cell of barley. Phytopathol. 58, 12 (1968).

BRADY, H. A., and F. A. PEEVY: Dimethyl sulfoxide fails to increase hardwood kills by herbicides. S. Weed Conf. Proc. 21, 218 (1968).

BRANTON, D.: Fracture faces of frozen membranes. Amer. J. Bot. 53, 603 (1966).

—— Membrane structure. Ann. Rev. Plant Physiol. 20, 209 (1969).

BRANTS, D. H.: Relation between ectodesmata and infection of leaves by C[14]-labeled tobacco mosaic virus. Virol. 26, 554 (1965).

BRAUN, H. J., and J. J. SAUTER: Phosphatase-Lokalisation in Phloembeckenzellen und Siebröhren der Dioscoreaceae und ihre mögliche Bedeutung für den aktiven Assimilattransport. Planta 60, 543 (1964).

BRECCIA, A., and A. ABBONDANZA: The role of general metabolites in the biosynthesis of natural products. II. The paraffin heptacosane. Z. Naturforsch. 22b, 50 (1967).

BRIAN, P. W.: Uptake and transport of systemic fungicides and bactericides. Agrochim. 11, 203 (1967).

BRIAN, R. C.: The effects of herbicides on biophysical processes in the plant. In: The physiology and biochemistry of herbicides, p. 357. London and New York: Academic Press (1964).

—— Action of plant growth regulators: IV. Adsorption of unsubstituted and 2,6-dichloro-aromatic acids to oat monolayers. Plant Physiol. 42, 1209 (1967 a).

—— The uptake and adsorption of diquat and paraquat by tomato, sugar beet, and cocksfoot. Ann. Applied Biol. 59, 91 (1967 b).

——, and N. D. CATTLIN: The surface structure of leaves of Chenopodium album L. Ann. Bot. 32, 609 (1968).

BRIQUET, M., G. SCHELLER, E. CRÈVECŒUR, and A. WIAUX: Étude de la migration de l'acide 2,4-dichlorophénoxybutyrique-1-^{14}C chez Phaseolus vulgaris L. Weed Research 8, 61 (1968).

BROVCHENKO, M. I.: O postuplenii sakharov iz mesofilla v provodyashchie puchki list'ev sakharnoi svekly. Fiziol. Rast. 12, 270 (1965).

BROWN, H. D., A. CHERRIE, and A. CASSENS: Environment and trichome morphogenesis in Nicotiana. Phytol. 7, 363 (1961).

BROWN, W. V., and Sr. C. JOHNSON: The fine structure of the grass guard cell. Amer. J. Bot. 49, 110 (1962).

BRUNSKILL, R. T.: Physical factors affecting the retention of spray droplets on leaf surfaces. 3rd Brit. Weed Control Conf. Proc. 2, 593 (1956).

BUCHANAN, G. A., and D. W. STANIFORTH: Surfactant toxicity to plant tissues. Weed Soc. Amer. Abstr., p. 44 (1966).

BUKOVAC, M. J.: Some factors affecting the adsorption of 3-chlorophenoxy-α-propionic acid by leaves of the peach. Amer. Soc. Hort. Sci. Proc. 87, 131 (1965).

——, and R. F. NORRIS: Significance of waxes in cuticular penetration of plant growth substances. Plant Physiol. **42**, S-48 (1967).

—— —— Foliar penetration of plant growth substances with special reference to binding by cuticular surfaces of pear leaves. Agrochim. **12**, 217 (1968).

BULLARD, R. W., and D. L. CAMPBELL: Evaluation of adhesives for foliar application of chemicals. Forest Sci. **14**, 39 (1968).

BURIAN, K.: Rhodamin-B-Färbbarkeit und Eigenfluoreszenz in Zellen heimischer und tropischer Orchideen. II. Primäre und pathologisch veränderte Eigenfluoreszenz. Protoplasma **58**, 561 (1964).

BURKOWICZ, A., and R. N. GOODMAN: Permeability alterations induced by apple leaves by virulent and avirulent strains of *Erwinia amylovora*. Phytopathol. **59**, 314 (1969).

BURNS, E. R., G. A. BUCHANAN, and A. E. HILTBOLD: Absorption and translocation of 2,4-D-2-C¹⁴ by wolftail (*Carex cherokeensis* Schwein). Weed Soc. Amer. Abstr., p. 42 (1967).

BUTTERFASS, TH.: Fluoroskopische Untersuchungen an Keulenhaaren von *Vicia faba* L. und *Phaseolus coccineus* L. Protoplasma **47**, 415 (1956).

BYSTROM, B. G., R. B. GLATER, F. M. SCOTT, and E. S. C. BOWLER: Leaf surface of *Beta vulgaris*—electron microscope study. Bot. Gaz. **129**, 133 (1968).

CALDERBANK, A.: The bipyridylium herbicides. Adv. Pest Control Research **8**, 127 (1968).

CASLEY-SMITH, J. R.: Some observations on the electron microscopy of lipids. J. Roy. Microsc. Soc. **87**, 463 (1967).

CHAKRABORTY, M. K., and J. A. WEYBREW: The chemistry of tobacco trichomes. Tobacco Sci. **7**, 122 (1963).

CHALLEN, S. B.: The contribution of surface characters to the wettability of leaves. J. Pharm. & Pharmacol. **12**, 307 (1960).

CHANG, C., and E. SIMON: The effect of dimethyl sulfoxide (DMSO) on cellular systems. Soc. Expt. Biol. Med. Proc. **128**, 60 (1968).

CHANG, F. Y., and W. H. VANDEN BORN: Translocation of dicamba in Canada thistle. Weed Sci. **16**, 176 (1968).

CHAPMAN, B.: Polystyrene replicas for scanning reflexion electron microscopy. Nature **216**, 1347 (1967).

CHOW, P. N., O. C. BURNSIDE, T. L. LAVY, and H. W. KNOCHE: Absorption, translocation, and metabolism of silvex in prickly pear. Weeds **14**, 38 (1966).

CHU, H., and T. C. TSO: Fatty acid composition in tobacco. I. Green tobacco plants. Plant Physiol. **43**, 428 (1968).

CLARKE, J.: Preparation of leaf epidermis for topographic study. Stain Technol. **35**, 35 (1960).

CLOR, M. A.: Translocation and metabolism of C¹⁴-labeled urea in cotton seedlings. Iraqui Sci. Soc. Proc. **5**, 27 (1962).

COLLANDER, R.: Cell membranes: Their resistance to penetration and their capacity for transport. In: Plant physiology a treatise, Vol. II. Plants in relation to water and solutes, pp. 3–102. New York and London: Academic Press (1959).

CORDS, H. P.: Root temperature and susceptibility to 2,4-D in three species. Weeds **14**, 121 (1966).

CORNS, W. G., and T. DAI: Effects of added surfactant on toxicity of picloram, 2,4-D and 2,4,5-T to *Populus tremuloides* Michx. and *P. balsamifera* L. saplings. Can. J. Plant Sci. **47**, 711 (1967).

CRAFTS, A. S.: The chemistry and mode of action of herbicides. New York: Interscience (1961).

—— Herbicide behavior in the plant. In: The physiology and biochemistry of herbicides, pp. 75–110. London and New York: Academic Press (1964).

—— Relation between food and herbicide transport. In: Proceedings of a symposium on the use of isotopes in weed research, pp. 3–7. Oct. 25–29, 1965, Vienna. Internat. Atomic Energy Agency: Vienna, Austria (1966).

—— Absorption and translocation of labeled tracers. Ann. N. Y. Acad. Sci. 144, 357 (1967).

——, and C. L. Foy: The chemical and physical nature of plant surfaces in relation to the use of pesticides and their residues. Residue Reviews 1, 112 (1962).

——, and S. Yamaguchi: The autoradiography of plant materials. Manual 35, Calif. Agr. Expt. Sta., Extension Service (1964).

Crisp, C. E.: Thè biopolymer cutin. Ph.D. Diss., Univ. Calif., Davis (1965).

——, D. A. Fisher, and D. E. Bayer: An electron microscopic study of Gilson diffusion patterns in *Plantago major*. W. Weed Control Conf. Research Progress Rept., p. 138 (1966).

Cunze, R.: Untersuchungen über die ökologische Bedeutung des Wachses im Wasserhaushalt der Pflanzen. Beih. Bot. Zbl. 42, 160 (1926).

Currier, H. B., and C. D. Dybing: Foliar penetration of herbicides—Review and present status. Weeds 7, 195 (1959).

——, and W. Eschrich: Callose: Synthesis, degradation, possible functions. Plant Physiol. (Suppl.) 39, xlvii (1964).

——, and C. Y. Shih: Sieve tubes and callose in *Elodea* leaves. Amer. J. Bot. 55, 145 (1968).

——, and D. H. Webster: Callose formation and subsequent disappearance: Studies in ultrasound stimulation. Plant Physiol. 39, 843 (1964).

Czaja, A. Th.: Die submikroskopische Struktur (Textur) der Aussenwände von Blattepidermen. Planta 57, 669 (1962).

Dainty, J.: Physiological aspects of ion transport in plant cells and tissues. In: The cellular functions of membrane transport, p. 41. New Jersey: Prentice-Hall (1964).

Daly, G. T.: Leaf surface wax in *Poa colensoi*. J. Expt. Bot. 15, 160 (1964).

Davenport, J. B.: Studies in the natural coating of apples. V. Unsaturated and minor saturated acids of the cuticle oil. Austral. J. Chem. 13, 411 (1960).

Davies, P. J., D. S. H. Drennan, J. D. Fryer, and K. Holly: The basis of the differential phytotoxicity of 4-hydroxy-3,5-di-iodobenzonitrile. I. The influence of spray retention and plant morphology. Weed Research 7, 220 (1967).

—— —— —— —— The basis of the differential phytotoxicity of 4-hydroxy-3,5-di-iodobenzonitrile. II. Uptake and translocation. Weed Research 8, 233 (1968).

Davis, F. S., R. W. Bovey, and M. G. Merkle: The role of light, concentration, and species in foliar uptake of herbicides in woody plants. Forest Sci. 14, 164 (1968 a).

—— —— Effect of paraquat and 2,4,5-T on the uptake and transport of picloram in woody plants. Weed Sci. 16, 336 (1968 b).

——, M. G. Merkle, and R. W. Bovey: Effect of moisture stress on the absorption and transport of herbicides in woody plants. Bot. Gaz. 129, 183 (1968 c).

Day, B. E.: The absorption and translocation of 2,4-dichlorophenoxyacetic acid by bean plants. Plant Physiol. 27, 143 (1952).

——, K. C. Barrons, S. N. Fertig, V. H. Freed, P. C. Hamm, C. B. Huffaker, D. L. Klingman, D. W. Staniforth, and R. P. Upchurch: Principles of plant and animal pest control. Vol. 2: Weed control. Nat. Acad. Sci., Washington, D. C. (1968).

Dilcher, D. L.: Revision of Eocene palms from southeastern North America based upon cuticular analysis. Amer. J. Bot. 55, 725 (1968).

Dimond, A. E.: Surface factors affecting the penetration of compounds into plants. In: Moderne Methoden der Pflanzenanalyse. Vol. 5, p. 368. Heidelberg: Springer-Verlag (1962).

—— Natural models for plant chemotherapy. Adv. Pest Control Research 6, 127 (1965).

Dolzmann, P.: Elektronenmikroskopische Untersuchungen an den Saughaaren von

Tillandsia usneoides (*Bromeliaceae*). I. Feinstruktur der Kuppelzelle. Planta 60, 461 (1964).

—— Elektronenmikroskopische Untersuchungen an den Saughaaren von *Tillandsia usneoides* (*Bromeliaceae*). II. Einige Beobachtungen zur Feinstruktur der Plasmodesmen. Planta 64, 76 (1965).

DONNALLEY, W. F., and S. K. RIES: Amitrole translocation in *Agropyron repens* increased by the addition of ammonium thiocyanate. Science 145, 497 (1964).

DONOHO, C. W. JR., A. E. MITCHELL, and M. J. BUKOVAC: The absorption and translocation of ring labeled C^{14}-naphthaleneacetic acid in the apple and peach. Amer. Soc. Hort. Sci. Proc. 78, 96 (1961).

DOROKHOV, B. L.: O vozmozhnosti vneust'ichnogo kutikulyarnogo fotosinteza u nekotorykh rastenii. Bot. Zh. 48, 893 (1963).

DOUGLAS, G.: The influence of size of spray droplets on the herbicidal activity of diquat and paraquat. Weed Research 8, 205 (1968).

DOUS, E.: Über Wachsausscheidungen bei Pflanzen; ein Studium mit dem Oberflächenmikroskop. Bot. Arch. 19, 461 (1927).

DRAWERT, H., and U. RÜEFFER-BOCK: Mikrospektralphotometrische Untersuchungen mit dem UMSP I (C. Zeiss) an Oberepidermen der Schuppenblätter von *Allium cepa* nach Neutralrotfärbung. Protoplasma 60, 435 (1965).

DREES, H.: Pesticide consumption over the course of 15 years. Gesunde Pflanz. 19, 244 (1967).

DULAEVA, O. N., E. A. POPOVA, I. BENESH, and M. E. DOLGAYA: Pogloshchenie mechenogo 6-benzilaminopurina vysechkami iz list'ev yachmenya i ego raspredelenie po fraktsiyam kletochnykh struktur. Fiziol. Rast. 14, 309 (1967).

DYSON, W. G., and G. A. HERBIN: Studies on plant cuticular waxes. IV. Leaf wax alkanes as a taxonomic discriminant for cypresses grown in Kenya. Phytochem. 7, 1339 (1968).

EBELING, W.: Analysis of the basic processes involved in the deposition, degradation, persistence, and effectiveness of pesticides. Residue Reviews 3, 35 (1963).

ECHLIN, P.: The use of the scanning reflection electron microscope in the study of plant and microbial material. J. Roy. Microsc. Soc. 88, 407 (1968).

EDDINGS, J. L., and A. L. BROWN: Absorption and translocation of foliar-applied iron. Plant Physiol. 42, 15 (1967).

EDGERTON, L. J., and W. J. GREENHALGH: Absorption, translocation and accumulation of labeled N-dimethylaminosuccinamic acid in apple tissues. Amer. Soc. Hort. Sci. Proc. 91, 25 (1967).

EGLINTON, G., and R. J. HAMILTON: Leaf epicuticular waxes. Science 156, 1322 (1967).

——, and D. H. HUNNEMAN: Gas chromatographic-mass spectrometric studies of long chain hydroxy acids. I. The constituent cutin acids of apple cuticle. Phytochem. 7, 313 (1968).

ENNIS, W. B., JR., R. E. WILLIAMSON, and K. P. DORSCHNER: Studies on spray retention by leaves of different plants. Weeds 1, 274 (1952).

EPTON, H. A. S., and B. J. DEVERALL: A biochemical difference between healthy bean leaves resistant and susceptible to the halo-blight disease caused by *Pseudomonas phaseolicola*. Ann. Applied Biol. 61, 255 (1968).

ESAU, K.: Plant anatomy, 2nd ed. New York: Wiley (1965 a).

—— Fixation images of sieve element plastids in Beta. Nat. Acad. Sci. Proc. 54, 429 (1965 b).

—— Minor veins in Beta leaves: Structure related to function. Amer. Phil. Soc. Proc. 111, 219 (1967).

——, H. B. CURRIER, and V. I. CHEADLE: Physiology of phloem. Ann. Rev. Plant Physiol. 8, 349 (1957).

EVANS, E., J. R. COX, J. W. H. TAYLOR, and R. L. RUNHAM: Some observations on size and biological activity of spray deposits produced by various formulations of copper oxychloride. Ann. Applied Biol. 58, 131 (1966).

EVANS, R. A., and R. E. ECKERT, JR.: Paraquat-surfactant combinations for control of downy brome. Weeds 13, 150 (1965).

FABBRICOTTI-OBERRAUCH, J.: Untersuchungen uber aktive Aufnahme der sauren Farbstoffe Cyanol, Orange G und Ponceau in Pflanzengeweben, II. Protoplasma 59, 531 (1965).

FINLAYSON, D. G., and H. R. MacCARTHY: The movement and persistence of insecticides in plant tissue. Residue Reviews 9, 114 (1965).

FISCHER, R. A.: Stomatal opening in isolated epidermal strips of Vicia faba. I. Response to light and to CO_2-free air. Plant Physiol. 43, 1947 (1968).

——, and T. C. HSIAO: Stomatal opening in isolated epidermal strips of Vicia faba. II. Responses to KCl concentration and the role of potassium absorption. Plant Physiol. 43, 1953 (1968).

FISHER, D. A., C. E. CRISP, and D. E. BAYER: Ultrastructural observations of nine species of conifer needle outer epidermal cell wall thin sections at various ages. Can. J. Bot., In press (1969).

FISHER, D. B.: An unusual layer of cells in the mesophyll of the soybean leaf. Bot. Gaz. 128, 215 (1967).

FISHER, D. J.: Phenolic compounds of the apple fruit cuticle. Ann. Rep. Agr. & Hort. Research Sta., Long Ashton, Bristol, p. 255 (1965).

FOGG, G. E.: Quantitative studies on the wetting of leaves by water. Roy. Soc. (London) Ser. B. Biol. Sci. Proc. 134, 503 (1947).

FORD, R. E., and C. G. L. FURMIDGE: Impact and spreading of spray drops on foliar surfaces. Monograph Soc. Chem. Ind. 25, 417 (1967 a).

—— —— Physiochemical studies on agricultural sprays. VIII. Viscosity and spray drop size of water-in-oil emulsions. J. Sci. Food Agr. 18, 419 (1967 b).

FOY, C. L.: Penetration and initial translocation of 2,2-dichloropropionic acid (dalapon) in individual leaves of Zea mays. Weeds 10, 35 (1962).

—— The influence of formulation, exposure time, and pH on the herbicidal action of dalapon foliar sprays tested on corn. Hilgardia 35, 125 (1963).

—— Review of herbicide penetration through plant surfaces. J. Agr. Food Chem. 12, 473 (1964 a).

—— Volatility and tracer studies with alkylamino-s-triazines. Weeds 12, 103 (1964 b).

——, and L. W. SMITH: Surface tension lowering, wettability of paraffin and corn leaf surfaces, and herbicidal enhancement of dalapon by seven surfactants. Weeds 13, 15 (1965).

—— —— The role of surfactants in modifying the activity of herbicidal sprays. In: Pesticidal formulations research. Physical and colloidal chemical aspects. Adv. Chem. Series 86, 55 (1969).

——, J. W. WHITWORTH, T. J. MUZIK, and H. B. CURRIER: The penetration, absorption, and translocation of herbicides. In: Environmental and other factors in the response of plants to herbicides. Oregon Agr. Expt. Sta. Tech. Bull. 100, p. 3 (1967).

FRANKE, W.: The entry of solutes into leaves by means of ectodesmata. In: Absorption and translocation of organic substances in plants. 7th Ann. Symp., Amer. Soc. Plant Physiol., S. Sect., p. 95 (1964).

—— Mechanisms of foliar penetration of solutions. Ann. Rev. Plant Physiol. 18, 281 (1967 a).

—— Ektodesmen und die peristomatäre Transpiration. Planta 73, 138 (1967 b).

—— Ektodesmen nehmen auch an der Transpiration der Pflanzen teil. Umsch. Wiss. Tech. 67, 699 (1967 c).

—— Ectodesmata and the cuticular penetration of leaves. Lecture, Soc. of Chem. Ind., London, April 2 (1968).

—— Ectodesmata in relation to binding sites for inorganic ions and urea on isolated cuticular membrane surfaces. Amer. J. Bot. 56, 432 (1969).

——, und M. PANIC: Ektodesmenstudien. IV. Über das Vorkommen von Ektodesmen in Gramineenblättern. Planta 77, 176 (1967).

Freed, V. H., and M. Montgomery: The effect of surfactants on foliar absorption of 3-amino-1,2,4-triazole. Weeds 6, 386 (1958).

Freiberg, S. R., and P. Payne: Foliar absorption of urea and urease activity in banana plants. Amer. Soc. Hort. Sci. Proc. 69, 226 (1957).

Frey-Wyssling, A.: Submicroscopic morphology of protoplasm. New York: Elsevier (1953).

—— Die Pflanzliche Zellwand. Heidelberg: Springer-Verlag (1959).

——, and K. Mühlethaler: Ultrastructural plant cytology. Amsterdam, London and New York: Elsevier (1965).

Fridvalszky, L.: Differentiation of cell wall ultrastructure in the hairs of Cucurbita pepo L. Acta Agron. Acad. Sci. Hung. 16, 273 (1967).

Furmidge, C. G. L.: Physico-chemical studies on agricultural sprays. VI. Survey of methods for measuring the wetting ability of spray formulations. J. Sci. Food Agr. 16, 134 (1965).

Garren, R., Jr.: Uptake and distribution of labeled dimethyl sulfoxide and its influence on nutritive element transport in plants. Ann. N. Y. Acad. Sci. 141, 127 (1967).

Geiger, D. R., and D. A. Cataldo: Leaf structure and translocation in sugar beet. Plant Physiol. 44, 45 (1969).

Gel'man, N. S.: O molekulyarnoi organizatsii biologicheskikh membran. USP Sovrem. Biol. 64, 379 (1967).

Géneau de Lamarlière, L.: Sur les membranes cutinisées des plantes aquatiques. Rev. Gén. Bot. 18, 289 (1906).

Gentner, W. A.: The influence of EPTC on external foliage wax deposition. Weeds 14, 27 (1966).

Glauert, A. M.: Electron microscopy of lipids and membranes. J. Roy. Microsc. Soc. 88, 49 (1968).

Goldsmith, M. H.: The transport of auxin. Ann. Rev. Plant Physiol. 19, 347 (1968).

Goodman, R. N.: The impact of antibiotics upon plant disease control. Adv. Pest Control Research 5, 1 (1962).

——, Z. Király, and M. Zaitlin: The biochemistry and physiology of infectious plant disease. Princeton, Toronto, London, Melbourne: D. van Nostrand (1967).

Gould, R. F.: Contact angle wettability, and adhesion. Adv. Chem. Series 43 (1964).

Gray, R. A.: Increasing the absorption of streptomycin by leaves and flowers with glycerol. Phytopathol. 46, 105 (1956).

Greenham, C. G.: Studies on herbicide contents in roots of skeleton weed (Chondrilla juncea L.) following leaf applications. Weed Research 8, 272 (1968).

Greenwald, H. L., E. B. Kice, M. Kenly, and J. Kelly: Determination of the distribution of nonionic surface active agents between water and iso-octane. Anal. Chem. 33, 465 (1961).

Grncarevic, M., and F. Radler: The effect of wax components on cuticular transpiration—Model experiments. Planta 75, 23 (1967).

Grun, P.: Ultrastructure of plant plasma and vacuolar membranes. J. Ultrastruct. Research 9, 198 (1963).

Gunning, B. E. S., J. S. Pate, and L. G. Briarty: Specialized "transfer cells" in minor veins of leaves and their possible significance in phloem translocation. J. Cell Biol. 37, C7 (1968).

Gunther, F. A., D. L. Lindgren, M. I. Elliot, and J. P. LaDue: Persistence of certain DDT deposits under field conditions. J. Econ. Entomol. 39, 624 (1946).

Günther, I., and G. B. Wortmann: Dust on the surface of leaves. J. Ultrastruct. Research 15, 522 (1966).

Guth, E. B. de: Distribucion y caracteristicas de las hojas de palmas flabeladas argentinas. Rev. Invest. Agropecuar Ser. 2, Biol. Prod. Veg. 3, 55 (1966).

Haas, R. H., and S. K. Lehman: Unpublished data (1969).

HABERLANDT, G. E., JR.: Physiological plant anatomy. London: Macmillan (1914).

HAILE-MARIAM, S. N., and S. H. WITTWER: Comparative permeability of Na, K, Rb, and Cs ions through isolated cuticular membranes of Euonymus japonicus. Plant Physiol. (Suppl.) 40, xii (1965).

HALEVY, A. H., and S. H. WITTWER: Foliar uptake and translocation of rubidium in bean plants as affected by root absorbed growth regulators. Planta 67, 375 (1965).

HALL, D. M.: A study of the surface wax deposits on apple fruit. Autral. J. Biol. Sci. 19, 1017 (1966).

—— The ultrastructure of wax deposits on plant leaf surfaces. II. Cuticular pores and wax formation. J. Ultrastruct. Research 17, 34 (1967 a).

—— Wax micro-channels in the epidermis of white clover. Science 158, 505 (1967 b).

HALLAM, N. D.: Sectioning and electron microscopy of Eucalyptus leaf waxes. Austral. J. Biol. Sci. 17, 587 (1964).

HALLIER, E.: Phytopathologie, Saftaufnahme der Blätter von aussen. Leipzig: Verlag V. Wilh. Engelmann (1868).

HAMMERTON, J. L.: Environmental factors and susceptibility to herbicides. Weeds 15, 330 (1967).

HANSZEN, K. J.: Vergleichende Untersuchungen über das Wachsen von Kohlehüllen unter dem Einfluss von Elektronenbeschuss, Temperung und Bestrahlung mit Elektromagnetischen Wellen. In: Proceedings European Regional Conf. Electron Microscopie De Nederlandse Vereniging voor Electronenmicroscopie, p. 673 (1961).

HART, W. J., and W. HURTT: The influence of DMSO on the phytotoxicity of several herbicides. 21st N.E. Weed Control Conf. Proc. Suppl. 21, 156 (1967).

HARTLEY, G. S.: Physics of foliar application in relation to formulation. 8th Brit. Weed Control Conf. Proc., p. 794 (1966).

HELDER, R. J.: Translocation in Vallisneria spiralis. In: Handbuch der Pflanzenphysiologie. XIII. Der Stofftransport in der Pflanze. Berlin, Heidelberg, New York: Springer-Verlag, p. 20 (1967).

HENDRICKS, S. B.: Salt transport across cell membranes. Amer. Sci. 52, 306 (1964).

HENDRYCY, K. E.: The effect of trichomes on transpiration and ion uptake in Verbascum thapsus L. Bios 39, 21 (1968).

HERBIN, G. A., and P. A. ROBINS: Studies on plant cuticular waxes—I. The chemotaxonomy of alkanes and alkenes of the genus Aloe (Liliaceae). Phytochem. 7, 239 (1968 a).

—— —— Studies on plant cuticular waxes—II. Alkanes from members of the genus Agave (Agavaceae) the genera Kalanchoe, Echeveria, Crassula and Sedum (Crassulaceae) and the genus Eucalyptus (Myrtaceae) with an examination of Hutchinson's subdivision of the Angiosperms into Herbaceae and Lignosae. Phytochem. 7, 257 (1968 b).

—— —— Studies on plant cuticular waxes—III. The leaf wax alkanes and ω-hydroxy acids of some members of the Cupressaceae and Pinaceae. Phytochem. 7, 1325 (1968 c).

HESLOP-HARRISON, Y., and J. HESLOP-HARRISON: Scanning electron microscopy of leaf surfaces. Proc. 2nd Ann. Scanning Electron Microsc. Symp. p. 117 (1969).

HOEHNE, W., and R. WASICKY: Contribuição estudo mecanismo de seletividade herbicida do 2,4-D, ou ácido 2,4-diclorofenoxi-acético. Rev. Quim. Farm., Rio de Janeiro 15, 141 (1950).

HOHLT, H. E., and D. N. MAYNARD: Studies on foliar leaching of cations from spinach. Amer. Soc. Hort. Sci. Proc. 90, 296 (1967).

HOKIN, L. W., and M. R. HOKIN: The chemistry of cell membranes. Scientific American 213, 78 (1965).

HOLLOWAY, P. J., and E. A. BAKER: Isolation of plant cuticles with zinc chloride-hydrochloric acid solution. Plant Physiol. 43, 1878 (1968).

136	Herbert M. Hull

Holly, K.: Herbicide selectivity in relation to formulation and application methods. In: The physiology and biochemistry of herbicides, p. 423. New York: Academic Press (1964).
Hölzl, J.: Eine Ultrastruktur in der Zellwand der Innenepidermis der Zwiebelschuppe von *Allium cepa*. Ektodesmen? Protoplasma **59**, 310 (1964).
——, und E. Bancher: Zur Physikochemie der vitalen Kern- und Plasmafluorochromierung von *Allium Cepa*—Epidermen. Protoplasma **64**, 157 (1967).
Honsell, E.: I coloranti vitali e i problemi della loro permeazione ed accumulo nella cellula vegetale. G. Bot. Ital. **72**, 287 (1965).
Horanic, G., and F. E. Gardner: An improved method of making epidermal imprints. Bot. Gaz. **128**, 144 (1967).
Horrocks, R. L.: Wax and the water vapour permeability of apple cuticle. Nature **203**, 547 (1964).
Horsfall, F., Jr., and R. C. Moore: The effect of spray additives and simulated rainwater on foliage curvature and thinning of apples by the sodium salt of naphthaleneacetic acid. Amer. Soc. Hort. Sci. Proc. **80**, 15 (1962).
Hoskins, W. M.: Some important properties of pesticide deposits on various surfaces. Residue Reviews **1**, 66 (1962).
Hsieh, J. J. C., and F. P. Hungate: [131]I diffusion through leaf cuticle. Amer. Soc. Plant Physiol., W. Sect. Meeting Abstr., p. 2 (1968).
Huber, B., E. Kinder, E. Obermüller, and H. Ziegenspeck: Spaltöffnungs-Dünnstschnitte im Elektronen-mikroskop. Protoplasma **46**, 380 (1956).
Huelin, F. E.: Studies in the natural coating of apples. IV. The nature of cutin. Austral. J. Biol. Sci. **12**, 175 (1959).
Hughes, E. E.: Hygroscopic additives to phenoxy herbicides for control of saltcedar. Weed Sci. **16**, 486 (1968).
Hughes, R. E., and V. H. Freed: The role of surfactants in the foliar absorption of indole-3-acetic acid (IAA). Weeds **9**, 54 (1961).
Hull, H. M.: Studies on herbicidal absorption and translocation in velvet mesquite seedlings. Weeds **4**, 22 (1956).
—— Anatomical studies demonstrating phloem inactivation and its dependency upon the interaction of concentrations of 2,4,5-trichlorophenoxyacetic acid and an anionic wetting agent. Plant Physiol. (Suppl.) **32**, 43 (1957).
—— The effect of day and night temperature on growth, foliar wax content, and cuticle development of velvet mesquite. Weeds **6**, 133 (1958).
—— Leaf structure as related to penetration of organic substances. In: Absorption and translocation of organic substances in plants. 7th Ann. Symp. Amer. Soc. Plant Physiol., S. Sect., p. 45 (1964 a).
—— Absorption and translocation of a 2,4,5-T ester as influenced by interactions of pH and concentrations of glycerol and latex in the carrier. W. Weed Control Conf. Research Progress Rept., p. 106 (1964 b).
—— Dimethyl sulfoxide as a herbicide carrier under different conditions of light intensity. W. Weed Control Conf. Proc. **20**, 12 (1965).
—— Uptake and movement of herbicides in plants. Herbic. Veg. Management Symp. Proc., p. 49 (1967 a).
—— BioScience **17**(1), cover photograph (1967 b).
——, G. E. Barrier, R. E. Frans, J. L. Hilton, E. L. Knake, D. E. Moreland, and W. H. Zick: Herbicide handbook of the Weed Society of America. Geneva, New York: Humphrey Press (1967).
——, and S. J. Shellhorn: Cuticle development in the field and greenhouse and its relationship to herbicidal response. W. Weed Control Conf. Research Progress Rept., p. 137 (1966).
—— —— Foliar absorption of 2,4,5-T from emulsions and straight oil carriers in combination with oil-soluble surfactants. W. Weed Control Conf. Research Progress Rept., p. 123 (1967).
—— —— Some herbicide-carrier interrelationships in the control of velvet mesquite. W. Soc. Weed Sci. Proc. **22**, 19 (1968).

——, and F. W. WENT: Life processes of plants as affected by air pollution. 2nd Nat. Air Pollution Symp. Proc., p. 122 (1952).

HÜLSBRUCH, M.: Zur Radialstreifung cutinisierter Epidermisaussenwände. I. Z. Pflanzenphysiol. 55, 181 (1966).

IDLE, D. B.: Scanning electron microscopy of leaf surface replicas and the measurement of stomatal aperture. Ann. Bot. 33, 75 (1969).

JACOBS, W. P., C. C. McCREADY, and D. J. OSBORNE: Transport of the auxin 2,4-dichlorophenoxyacetic acid through abscission zones, pulvini, and petiols of *Phaseolus vulgaris*. Plant Physiol. 41, 725 (1966).

JACOBSON, J. S.: Accumulation and translocation of fluoride in plants. Amer. Soc. Plant Physiol. Proc. 1966, xxxv (1966).

JACOBY, B., and J. DAGAN: A comparison of two methods of investigating sodium uptake by bean-leaf cells and the vitality of isolated leaf-cells. Protoplasma 64, 325 (1967).

JAMIESON, G. R., and E. H. REID: Analysis of oils and fats by gas chromatography. V. Fatty acid composition of the leaf lipids of *Myosotis scorpioides*. J. Sci. Food Agr. 19, 628 (1968).

JANSEN, L. L.: Surfactant enhancement of herbicide entry. Weeds 12, 251 (1964).

—— Effects of structural variations in ionic surfactants on phytotoxicity and physical-chemical properties of aqueous sprays of several herbicides. Weeds 13, 117 (1965 a).

—— Herbicidal and surfactant properties of long-chain alkylamine salts of 2,4-D in water and oil sprays. Weeds 13, 123 (1965 b).

—— Selective modification of postemergence toxicity of atrazine and diuron by surfactant and oil adjuvants. Weed Soc. Amer. Abstr., p. 34 (1966).

JENNER, C. F., P. F. SAUNDERS, and G. E. BLACKMAN: The uptake of growth substances. 10. The accumulation of phenoxyacetic acid and 2,4-dichlorophenoxyacetic acid by segments of *Avena* mesocotyl. J. Expt. Bot. 19, 333 (1968 a).

—— —— —— The uptake of growth substances. 11. Variations in the accumulation of substituted phenoxyacetic acids of differing physiological activity by segments of *Avena* mesocotyl. J. Expt. Bot. 19, 353 (1968 b).

JENNINGS, D. H.: The absorption of solutes by plant cells. Edinburgh: Oliver and Boyd (1963).

JENSEN, W. A.: Botanical histochemistry. San Francisco: W. H. Freeman (1962).

JOHNSTON, H. W., and T. SPROSTON, JR.: The inhibition of fungus infection pegs in *Ginkgo biloba*. Phytopathol.. 55, 225 (1965).

JUNIPER, B. E.: Growth, development, and effect of the environment on the ultrastructure of plant surfaces. J. Linn. Soc. (Bot.) 56, 413 (1960 a).

—— Leaf surfaces under the electron microscope. 4th Internat. Conf. Electron Microsc. Proc., p. 489 (1960 b).

JYUNG, W. H., and S. H. WITTWER: Pathways and mechanisms for foliar absorption of mineral nutrients. Agr. Sci. Rev. 3, 26 (1965).

KAMIMURA, S., and R. N. GOODMAN: Penetration of excised apple cuticle membranes. II. Diffusion of radioactive model compounds and antibiotics and an analysis of membrane properties. Phytopathol. Z. 51, 324 (1964 a).

—— —— Influence of foliar characteristics on the absorption of a radioactive model compound by apple leaves. Physiol. Plant. 17, 805 (1964 b).

KANNAN, S., and S. H. WITTWER: Absorption of iron by enzymically isolated leaf cells. Physiol. Plant. 20, 911 (1967 a).

—— —— Ion absorption by leaves. Naturwiss. 54, 373 (1967 b).

—— Penetration of iron and some organic substances through isolated cuticular membranes. Plant Physiol. 44, 517 (1969).

KEIL, H. L.: DMSO shows great promise as carrier of agricultural toxicants. Agr. Chemicals 20(4), 23 (1965).

—— Enhanced bacterial spot control of peach when dimethyl sulfoxide is combined with sprays of oxytetracycline. In: Biological actions of dimethyl sulfoxide. Ann. N. Y. Acad. Sci. 141, 131 (1967).

KEMPEN, H., and D. E. BAYER: MSMA absorption, translocation and toxicity. Weed Sci. Soc. Amer. Abstr., No. 160 (1969).

KERR, A., and N. T. FLENTJE: Host infection in *Pellicularia filamentosa* controlled by chemical stimuli. Nature 179, 204 (1957).

KETELLAPPER, H. J.: Stomatal physiology. Ann. Rev. Plant Physiol. 14, 249 (1963).

KIRKWOOD, R. C., M. M. ROBERTSON, and J. E. SMITH: Differential absorption as a factor influencing the selective toxicity of MCPA and MCPB. In: Proceedings Symp. Use of Isotopes in Weed Research, p. 47 Oct. 25–29, 1965, Vienna. Internat. Atomic Energy Agency: Vienna, Austria (1966).

KOLATTUKUDY, P. E.: Biosynthesis of surface lipids. Science 159, 498 (1968 a).

—— Further evidence for an elongation-decarboxylation mechanism in the biosynthesis of paraffins in leaves. Plant Physiol. 43, 375 (1968 b).

—— Species specificity in the biosynthesis of branched paraffins in leaves. Plant Physiol. 43, 1423 (1968 c).

KOZEL, P. C., and H. B. TUKEY, JR.: Loss of gibberellins by leaching from stems and foliage of *Chrysanthemum morifolium* 'Princess Anne.' Amer. J. Bot. 55, 1184 (1968).

KREGER, D. R.: Wax. In: Handbuch der Pflanzenphysiologie, Bd. X, p. 249. Berlin-Göttingen-Heidelberg: Springer-Verlag (1958).

KREIDEMANN, P.: Sugar uptake and translocation in the castor bean seedling. III. An autoradiographic study of the absorption pathway. Planta 73,, 175 (1967).

KURSANOV, A. L.: Transport organicheskikh veshchestv v rasteniyakh. Izv. Akad. Nauk. SSSR Ser. Biol. 1, 3 (1967).

——, M. I. BROVCHENKO, and E. P. BUTENKO: Prevrashcheniya sakharov, absorbirovannykh tkanyami listovoi plastinki sakharnoi svekly. Fiziol. Rast. 14, 813 (1967).

KURTZ, E. B., JR.: The relation of the characteristics and yield of wax to plant age. Plant Physiol. 25, 269 (1950).

LADONIN, V. F.: O vliyanii efirov 2,4-D na azotnyi obmen rastenii. Akad. Nauk. Beloruss. SSR: Minsk, p. 70 (1961).

LANG, R. T.: Concerning the morphology of isolated plant cuticles. New Phytol. 68, 423 (1969).

LAPHAM, V. T.: The effectiveness of some dimethylsulfoxide-herbicide combinations. 19th Ann. Meeting S. Weed Conf. Proc., p. 438 (1966).

LASETER, J. L., D. J. WEBER, and J. ORÓ: Characterization of cabbage leaf lipids: n-alkanes, ketone, and fatty acids. Phytochem. 7, 1005 (1968 a).

——, J. WEETE, and D. J. WEBER: Alkanes, fatty acid methyl esters, and free fatty acids in surface wax of *Ustilago maydis*. Phytochem. 7, 1177 (1968 b).

LATIES, G. G.: Physiological aspects of membrane function in plant cells during development. In: Cellular membranes in development, p. 299. New York and London: Academic Press (1964).

—— Dual mechanisms of salt uptake in relation to compartmentation and long-distance transport. Ann. Rev. Plant Physiol. 20, 89 (1969).

LÄUCHLI, A., und U. LÜTTGE: Untersuchung der Kinetik der Ionenaufnahme in das Cytoplasma von *Mnium*-Blattzellen mit Hilfe der Mikroautoradiographie und der Röntgen-Mikrosonde. Planta 83, 80 (1968).

LAUTENSCHLAGER-FLEURY, D.: Über die Ultravioletdurchlassigkeit von Blattepidermen. Ber. Schweiz. Bot. Ges. 65, 343 (1955).

LAYNE, R. E. C.: Foliar trichomes and their importance as infection sites for *Corynebacterium michiganese* on tomato. Phytopathol. 57, 981 (1967).

LEAKE, C. D.: Biological actions of dimethyl sulfoxide. Ann. N. Y. Acad. Sci. 141, 1 (1967).

LEDBETTER, M. C.: Observations on membranes in plant cells fixed with OsO₄. 5th Internat. Congress for Electron Microsc., p. W-10. New York and London: Academic Press (1962).

LEHMAN, S. K., R. H. HAAS, and L. F. BOUSE: Spray distribution characteristics of hydroxy-ethyl cellulose sprays applied from fixed-wing aircraft. Weed Sci. Soc. Amer. Abstr., p. 120 (1968).

—— ——, and E. D. ROBISON: Preliminary evaluation of droplet size and distribution of water-in-oil emulsion sprays applied with the bifluid system. 17th S. Weed Conf. Proc., p. 253 (1964).

LEIGH, J. H., and J. W. MATTHEWS: An electron microscope study of the wax bloom on leaves of certain love grasses, (Eragrostis curvula(Schrad.) Nees). Austral. J. Bot. 11, 62 (1963).

LEONARD, C. D.: Use of dimethyl sulfoxide as a carrier for iron in nutritional foliar sprays applied to citrus. In: Biological actions of dimethyl sulfoxide. Ann. N. Y. Acad. Sci. 141, 148 (1967).

LEONARD, O. A., and R. K. GLENN: Translocation of herbicides in detached bean leaves. Weed Sci. 16, 352 (1968 a).

—— —— Translocation of assimilates and phosphate in detached bean leaves. Plant Physiol. 43, 1380 (1968 b).

——, and D. L. KING: Vein loading and transport in detached leaves. Plant Physiol. 43, 460 (1968).

——, L. A. LIDER, and R. K. GLENN: Absorption and translocation of herbicides by Thompson Seedless (Sultanina) grape, Vitis vinifera L. Weed Research 6, 37 (1966).

LEOPOLD, A. C., P. VAN SCHAIK, and M. NEAL: Molecular structure and herbicide adsorption. Weeds 8, 48 (1960).

LE VAILLANT, M., and M. R. GORENFLOT: Développement hétéroblastique, épiderme et trichomes de Crupina vulgaris Pers. var. brachypappa Beauv. C. R. Acad. Sci. Sér. D. 267, 177 (1968).

LEVI, E.: Chemical control of Typha angustifolia L. var. brownii. Weeds 8, 128 (1960).

—— The distribution of mineral elements following leaf and root uptake. Physiol. Plant. 21, 213 (1968).

LEWIS, S., and R. N. GOODMAN: Mode of penetration and movement of fire blight bacteria in apple leaf and stem tissue. Phytopathol. 55, 719 (1965).

LEYTON, L., and I. P. ARMITAGE: Cuticle structure and water relations of the needles of Pinus radiata (D. Don). New Phytol. 67, 31 (1968).

——, and B. E. JUNIPER: Cuticle structure and water relations of pine needles. Nature 198, 770 (1963).

LIKHOLAT, T. V.: Rol' sul'fata ammoniya v povyshenii gerbitsidnoi aktivnosti 2,4-D v bor'be s sornyakami. In: Okhrana prirody Tsentralno-chernozemnoi polosy. Voronezh 4, 205 (1962).

LINSCOTT, D. L., and R. D. HAGIN: Effects of two environmental factors on removal of 2,4-DB from forage. Weed Sci. 16, 114 (1968).

LINSER, H.: The design of herbicides. In: The physiology and biochemistry of herbicides, p. 483. London and New York: Academic Press (1964).

LINSKENS, H. F.: Das Relief der Blattoberfläche. Planta 68, 1 (1966).

——, and A. GELISSEN: Die Natur der Rauhschaligkeit bei Früchten der Apfelsorte "Golden Delicious." Phytopathol. Zeit. 57, 1 (1966).

——, W. HEINEN, and A. L. STOFFERS: Cuticula of leaves and the residue problem. Residue Reviews 8, 136 (1965).

——, and H. KROES: Interference microscopy of the pattern of leaf surfaces. Nature 210, 968 (1966).

LITTLE, E. C. S., and G. E. BLACKMAN: The movement of growth regulators in plants. III. Comparative studies of transport in Phaseolus vulgaris. New Phytol. 62, 173 (1963).

LITZ, R. E., and W. C. KIMMINS: Plasmodesmata between guard cells and accessory cells. Can. J. Bot. 46, 1603 (1968).

LOCKE, M.: Cellular membranes in development. 22nd Symp. Soc. Study Development and Growth. New York and London: Academic Press (1964).

LOCKHART, J. A., and U. B. FRANZGROTE: The effects of ultraviolet radiation on plants. Encyclopedia Plant Physiol. 16, 532 (1961).

LOEWENSTEIN, W. R.: Biological membranes: Recent progress. Ann. N. Y. Acad. Sci. 137, 403 (1966).

LÓPEZ AROCHA, M., and D. RINCÓN: Absorption and translocation of simazine and ametryne applied to leaves and roots: Movement of these compounds in the soil. Mem. 6ᵃˢ J. Agron. Maracaibo 1, 14 (1966).

LÓPEZ-SÁEZ, J. F., G. GIMÉNEZ-MARTÍN, and M. C. RISUEÑO: Fine structure of the plasmodesm. Protoplasma 61, 81 (1966).

LUHAN, M.: Die Epidermis von Agropyron repens. Protoplasma 56, 645 (1963).

LUKE, H. H., and T. E. FREEMAN: Effects of cytokinins on the absorption of victorin. Phytopathol. 58, 258 (1968).

LUND-HØIE, K., and D. E. BAYER: Absorption, translocation and metabolism of 3-amino-1,2,4-triazole in Pinus ponderosa and Abies concolor. Physiol. Plant. 21, 196 (1968).

LÜTTGE, U.: Aktiver Transport (Kurzstreckentransport bei Pflanzen). Protoplasmatologia. Handbuch der Protoplasmaforschung, Vol. VIII/7/b. Wien and New York: Springer-Verlag (1968).

——, and K. BAUER: Evaluation of ion uptake isotherms and analysis of individual fluxes of ions. Planta 80, 52 (1968).

——, and G. KRAPF: Die Ultrastruktur der Blattzellen junger und alter Mnium-Sprosse und ihr Zusammenhang mit der Ionenaufnahme. Planta 81, 132 (1968).

MACHADO, R. D.: Observaçoes sôbre a fôlha e revestimento ceroso de Syagrus coronata (Mart.) Becc. Arq. Jard. Bot., Rio de Janeiro 16, 117 (1958).

MADER, H.: Cutin. In: Handbuch der Pflanzenphysiologie, Bd. X, p. 270. Berlin-Göttingen-Heidelberg: Springer-Verlag (1958).

MAERCKER, U.: Über vermeintliche Poren in Epidermisaussenwänden von Cocculus laurifolius und Camelia japonica. Z. Pflanzenphysiol. 53, 86 (1965 a).

—— Beiträge zur Histochemie der Schliesszellen. Protoplasma 60, 173 (1965 b).

MAESTRI, M.: Structural and functional effects of endothall on plants. Ph.D. Diss., Univ. Calif., Davis (1967).

MAGALHÃES, A. C., and F. M. ASHTON: Effect of dicamba on oxygen uptake and cell membrane permeability in leaf tissue of Cyperus rotundus L. Weed Research 9, 48 (1969).

—— ——, and C. L. FOY: Translocation and fate of dicamba in purple nutsedge. Weed Sci. 16, 240 (1968).

MAHAN, J. N., D. L. FOWLER, and H. H. SHEPARD: The pesticide review—1968. U. S. Department of Agriculture, Agr. Stabil. & Conserv. Service Publ. (1968).

MAIER, U.: Dendritenartige Strukturen in der Cuticularschicht von Lilium candidum. Protoplasma 65, 243 (1968).

MALCOLM, C. V., L. H. STOLZY, and C. R. JENSEN: Effect of artificial leaf coatings on foliar chloride uptake during sprinkler irrigation. Hilgardia 39, 69 (1968).

MALISAUSKIENE, V.: Dependence of the effect of the herbicide 2,4-D on the structure and conditions of the outer tissues of plants. Liet. TSR Mokslu. Akad. Darb. (Ser. C) 1, 31 (1964).

MARCHANT, R., and A. W. ROBARDS: Membrane systems associated with the plasmalemma of plant cells. Ann. Bot. 32, 457 (1968).

MARTIN, D. J.: Features on plant cuticle. An aid to the analysis of the natural diet of grazing animals, with especial reference to Scottish Hill sheep. Trans. Bot. Soc. Edin. 36, 278 (1955).

MARTIN, J. T.: The plant cuticle: Its structure and relation to spraying problems. Ann. Rept. East Malling Research Sta. 1960, p. 40 (1961).
—— Role of cuticle in the defense against plant disease. Ann. Rev. Phytopathol. 2, 81 (1964).
—— The cuticle of plants. Nat. Agr. Adv. Service Quart. Rev. 72, 139 (1966).
——, E. A. BAKER, and R. J. W. BYRDE: The fungitoxicities of cuticular and cellular components of citrus lime leaves. Ann. Applied Biol. 57, 491 (1966).
——, R. F. BATT, and R. T. BURCHILL: Fungistatic properties of apple leaf wax. Nature 180, 796 (1957).
——, and D. J. FISHER: The surface structure of plant roots. Ann. Rept. Agr. & Hort. Research Sta., Long Ashton, Bristol, p. 251 (1965).
——, and E. SOMERS: Solubilization of copper by leaves and water-soluble acids from leaf wax. Nature 180, 797 (1957).
MASUDA, Y.: Über den Einfluss von Auxin auf die Stoffpermeabilität des Protoplasmas. I. Mitteilung. Bot. Mag. Tokyo 66, 256 (1953).
MATSUDA, K.: The biosynthesis of waxes in plants. Ph.D. Diss. Univ. Ariz. (1962).
McCREADY, C. C.: Translocation of growth regulators. Ann. Rev. Plant Physiol. 17, 283 (1966).
McNAIR, J. B.: Some properties of plant waxes in relation to climate of habitat. Amer. J. Bot. 18, 518 (1931).
McNAIRN, R. B., and H. B. CURRIER: Translocation blockage by sieve plate callose. Planta 82, 369 (1968).
MECKLENBURG, R. A., H. B. TUKEY, JR., and J. V. MORGAN: A mechanism for the leaching of calcium from foliage. Plant Physiol. 41, 610 (1966).
MERCER, F.: The submicroscopic structure of the cell. Ann. Rev. Plant Physiol. 11, 1 (1960).
MERKLE, M. G., and F. S. DAVIS: Effect of moisture stress on absorption and movement of picloram and 2,4,5-T in beans. Weeds 15, 10 (1967).
MERTEN, D., and W. BUCHHEIM: Fixation of radioactive material as influenced by the microstructure of the plant surface. Proc. Internat. Symp. Radioecological Conc. Processes, p. 485. London and New York: Pergamon Press (1967).
METCALFE, C. R., and L. CHALK: Anatomy of the dicotyledons, vol. II. Oxford: Oxford University Press (1950).
MICHIE, M. J., and W. W. REID: Biosynthesis of complex terpenes in the leaf cuticle and trichomes of Nicotiana tabacum. Nature 218, 578 (1968).
MILBORROW, B. V., and D. A. WILLIAMS: A re-examination of the penetration of Nitella cells by non-electrolytes. Physiol. Plant. 21, 902 (1968).
MILLER, C. S., and M. M. ABOUL-ELA: Absorption of S,S,S-tributylphosphorotrithioate by cotton leaves. J. Agr. Food Chem. 16, 946 (1968).
MINSHALL, W. H.: The effect of environment on the comparative resistance of petiole parenchyma to petroleum oils. Weeds 9, 356 (1961).
MIROSLAVOV, E. A.: The content of peroxidase, ascorbic acid and sucrose in the trichomes of certain plants. Bot. Zh. 44, 550 (1959).
—— Nekotorye cherty kseromorfnogo stroeniya epidermisa lista ryada zlakov. Bot. Zh. 47, 1339 (1962).
—— Elektronnomikroscopicheskoe issledovanie ust'its lista rzhi. Bot. Zh. 51, 446 (1966).
MITCHELL, J. W.: Progress in research on absorption, translocation, and exudation of biologically active compounds in plants. In: Perspectives of biochemical plant pathology. Conn. Agr. Expt. Sta. Bull. 663, 49 (1963).
——, and P. J. LINDER: Absorption, translocation, exudation, and metabolism of plant growth-regulating substances in relation to residues. Residue Reviews 2, 51 (1963).
——, and G. A. LIVINGSTON: Methods of studying plant hormones and growth-regulating substances. Agr. Handbook No. 336, Agr. Research Service, U. S. Department of Agriculture (1968).

——, B. C. SMALE, and R. L. METCALF: Absorption and translocation of regulators and compounds used to control plant diseases and insects. Adv. Pest Control Research 3, 359 (1960).

MOKHNACHEV, I. G., V. P. PISKLOV, and L. A. DULAN: Sostav parafinovykh uglevodorodov tabaka. Prikl. Biokhim. Mikrobiol. 3, 246 (1967).

MORELAND, D. E., G. H. EGLEY, A. D. WORSHAM, and T. J. MONACO: Regulation of plant growth by constituents from higher plants. Adv. Chem. Series 53, 112 (1966).

MORESHET, S,. D. KOLLER, and G. STANHILL: The partitioning of resistances to gaseous diffusion in the leaf epidermis and the boundary layer. Ann. Bot. 32, 695 (1968).

MORTON, H. L.: Influence of temperature and humidity on foliar absorption, translocation, and metabolism of 2,4,5-T by mesquite seedlings. Weeds 14, 136 (1966).

——, and J. A. COOMBS: Influence of surfactants on phytotoxicity of a picloram-2,4,5-T spray on three woody plants. Weed Sci. Soc. Amer. Abstr., p. 65 (1969).

——, F. S. DAVIS, and M. G. MERKLE: Radioisotopic and gas chromatographic methods for measuring absorption and translocation of 2,4,5-T by mesquite. Weed Sci. 16, 88 (1968).

MOTHES, K., and L. ENGELBRECHT: Kinetin-induced directed transport of substances in excised leaves in the dark. Phytochem. 1, 58 (1961).

MUELLER, S.: The taxonomic significance of cuticular patterns within the genus *Vaccinium* (Ericaceae). Amer. J. Bot. 53, 633 (1966).

MÜHLETHALER, K.: Plant cell walls. In: The cell, vol. II, p. 85. New York: Academic Press (1961).

—— Ultrastructure and formation of plant cell walls. Ann. Rev. Plant Physiol. 18, 1 (1967).

MULLER, C. H.: Inhibitory terpenes volatalized from *Salvia* shrubs. Bull. Torrey Bot. Club 92, 38 (1965).

MÜLLER, K., and A. C. LEOPOLD: Correlative aging and transport of P^{32} in corn leaves under the influence of kinetin. Planta 68, 167 (1966 a).

—— —— The mechanism of kinetin-induced transport in corn leaves. Planta 68, 186 (1966 b).

MUSSELL, H. W., D. J. MORRÉ, and R. J. GREEN: Response of plant tissues to dimethyl sulfoxide (DMSO). Plant Physiol. (Suppl.) 40, xiii (1965).

—— —— —— Acceleration of bean leaf abscission by 2,4-dichlorophenoxyacetic acid applied in dimethylsulfoxide. Can. J. Plant Sci. 47, 635 (1967).

NAKAMURA, H.: Effect of a surfactant on the herbicidal activity of s-triazine herbicides in foliar sprays. Weed Research, Japan 6, 48 (1967).

NAKATA, S., and A. C. LEOPOLD: Radioautographic study of translocation in bean leaves. Amer. J. Bot. 54, 769 (1967).

NAPP-ZINN, K.: Über Ektodesmen und verwandte Erscheinungen. Ber. Deut. Bot. Ges. 74, 62 (1961).

NELSON, C. D., and P. R. GORHAM: Uptake and translocation of C^{14}-labelled sugars applied to primary leaves of soybean seedlings. Can. J. Bot. 35, 339 (1957).

NETHERY, A. A.: Mitotic inhibition by surface-active agents. Amer. J. Bot. 54, 646 (1967 a).

—— Inhibition of mitosis by surfactants. Cytol. 32, 321 (1967 b).

——, and W. HURTT: Dimethyl sulfoxide-induced modifications of growth in *Phaseolus vulgaris* L. 'Black Valentine' and 'Red Kidney.' Amer. J. Bot. 54, 646 (1967).

NEUMANN, ST., und F. JACOB: Aufnahme von α-Aminoisobuttersäure durch die Blätter von *Vicia faba* L. Naturwiss. 55, 89 (1968).

NOBLE, W. M.: Air pollution effects. Pattern of damage produced on vegetation by smog. J. Agr. Food Chem. 3, 330 (1955).

Norris, D. M.: Systemic insecticides in trees. Ann. Rev. Entomol. 12, 127 (1967).

Norris, L. A.: The physiological and biochemical basis of selective herbicide action. Herbic. Veg. Management Symp. Proc., p. 56 (1967).

——, and V. H. Freed: Dimethyl sulfoxide as an absorption and translocation aid. W. Weed Control Conf. Research Progress Rept., p. 85 (1963).

—— —— The absorption and translocation characteristics of several phenoxy-alkyl acid herbicides in bigleaf maple. Weed Research 6, 203 (1966).

Norris, R. F., and M. J. Bukovac: Structure of the pear leaf cuticle with special reference to cuticular penetration. Amer. J. Bot. 55, 975 (1968).

Northcote, D. H.: Structure and function of plant-cell membranes. Brit. Med. Bull. 24, 107 (1968).

O'Brien, T. P.: Note on an unusual structure in the outer epidermal wall of the —— Observations on the fine structure of the oat coleoptile. I. The epidermal cells of the extreme apex. Protoplasma 63, 385 (1967).

Oland, K., and T. B. Opland: Uptake of magnesium by apple leaves. Physiol. Plant. 9, 401 (1956).

Orgell, W. H.: Sorptive properties of plant cuticle. Iowa Acad. Sci. Proc. 64, 189 (1957).

Ormrod, D. J., and A. J. Renney: A survey of weed leaf stomata and trichomes. Can. J. Plant Sci. 48, 197 (1968).

Osborne, D. J., and M. Hallaway: The role of auxins in the control of leaf senescence. Some effects of local applications of 2,4-dichlorophenoxyacetic acid on carbon and nitrogen metabolism. 4th Internat. Conf. Plant Growth Regulators Proc., p. 329 (1961).

Osmond, C. B.: Ion absorption in Atriplex leaf tissue. I. Absorption by mesophyll cells. Austral. J. Biol. Sci. 21, 1119 (1968).

——, and G. G. Laties: Interpretation of the dual isotherm for ion absorption in beet tissue. Plant Physiol. 43, 747 (1968).

Overbeek, J. van: Absorption and translocation of plant regulators. Ann. Rev. Plant Physiol. 7, 355 (1956).

Pallas, J. E., Jr., and G. G. Williams: Foliar absorption and translocation of P^{32} and 2,4-dichlorophenoxyacetic acid as affected by soil-moisture tension. Bot. Gaz. 123, 175 (1962).

Palmer, R. D., and W. B. Ennis, Jr.: Periderm formation in hypocotyl of Gossypium hirsutum L. and its effect upon penetration of an herbicidal oil. Weeds 8, 89 (1960).

Pardee, A. B.: Membrane transport proteins. Science 162, 632 (1968).

Parr, J. F., and A. G. Norman: Considerations in the use of surfactants in plant systems: A review. Bot. Gaz. 126, 86 (1965).

Pennington, L. R., and L. C. Erickson: Preliminary trials to determine the activating effects of DMSO on the phytotoxicity of dicamba and 2,4-D to mullein. W. Weed Control Conf. Research Progress Rept., p. 102 (1966).

Penot, M.: Action du fluorure de sodium sur l'absorption des phosphates par des disques de feuille de Nicotiana tabacum: Inhibition, stimulation. C. R. Hebd. Seances Acad. Sci. Ser. D Sci. Nature 264, 926 (1967).

Petetin, C. A.: Identificacion de plantulas y rebrotes de malezas Compuestas (Cardos y abrepunos) y Umbeliferas. Rev. Invest. Agropecuar., Ser. 2 Biol. Prod. Veg. 1, 133 (1964).

Phillips, R. L., and M. J. Bukovac: Influence of root temperature on absorption of foliar applied radiophosphorus and radiocalcium. Amer. Soc. Hort. Sci. Proc. 90, 555 (1967).

Pickering, E. R.: Foliar penetration pathways of 2,4-D, monuron, and dalapon as revealed by historadioautography. Ph.D. Diss., Univ. Calif., Davis (1965).

Pietri-Tonelli, P. de: Penetration and translocation of Rogor applied to plants. Adv. Pest Control Research 6, 31 (1965).

PLAUT, Z., and L. REINHOLD: The effect of water stress on the movement of [¹⁴C] sucrose and of tritiated water within the supply leaf of young bean plants. Austral. J. Biol. Sci. **20**, 297 (1967).

PLUMMER, G. L., and J. B. KETHLEY: Foliar absorption of amino acids, peptides, and other nutrients by the pitcher plant, *Sarracenia flava*. Bot. Gaz. **125**, 245 (1964).

POSSINGHAM, J. V., T. C. CHAMBERS, F. RADLER, and M. GRNCAREVIC: Cuticular transpiration and wax structure and composition of leaves and fruits of *Vitis vinifera*. Austral. J. Biol. Sci. **20**, 1149 (1967).

PRASAD, R., C. L. FOY, and A. S. CRAFTS: Role of relative humidity and solution additives on the foliar absorption and translocation of radio-labeled 2,2-dichloropropionic acid (Dalapon). Plant Physiol. (Suppl.) **37**, xiii (1962).

—— —— —— Effects of relative humidity on absorption and translocation of foliarly applied dalapon. Weeds **15**, 149 (1967).

PREECE, T. F., G. BARNES, and J. M. BAYLEY: Junctions between epidermal cells as sites of appressorium formation by plant pathogenic fungi. Plant Pathol. **16**, 117 (1967).

PURDY, S. J., and E. V. TRUTER: Taxonomic significance of surface lipids of plants. Nature **190**, 554 (1961).

QUASTEL, J. H.: Molecular transport at cell membranes. Roy. Soc. Series B. Biol. Sci. **163**, 169 (1965).

QUIMBY, P. C., JR., and J. D. NALEWAJA: Uptake of dicamba-C¹⁴ by leaf sections of wheat and wild buckwheat. Weed Sci. Soc. Amer. Abstr., p. 15 (1968).

RADLER, F.: Reduction of loss of moisture by the cuticle wax components of grapes. Nature **207**, 1002 (1965).

RADUNZ, A.: Über die Fettsauren in Blättern und Chloroplasten von *Antirrhinum majus* in Abhängigkeit von der Entwicklung. Flora, Abstr. A. **157**, 131 (1966).

RAINS, D. W.: Kinetics and energetics of light-enhanced potassium absorption by corn leaf tissue. Plant Physiol. **43**, 394 (1968).

RAM, C. S. V.: Tea leaf wax as a stimulant and fungistat of spore germination. Current Sci. **31**, 428 (1962).

RAO, A. N.: Reticulate cuticle on leaf epidermis in *Hevea brasiliensis* Muell. Nature **197**, 1125 (1963).

RICHTER, H.: Die Reaktion hochpermeabler Pflanzenzellen auf drei Gefrierschutzstoffe (Glyzerin, Äthylenglykol, Dimethylsulfoxid). Protoplasma **65**, 155 (1968).

RINGOET, A., R. V. RECHENMANN, and H. VEEN: Calcium movement in oat leaves measured by semi-conductor detectors. Radiat. Bot. **7**, 81 (1967).

——, G. SAUER, and A. J. GIELINK: Phloem transport of calcium in oat leaves. Planta **80**, 15 (1968).

ROBARDS, A. W.: Desmotubule—A plasmodesmatal substructure. Nature **218**, 784 (1968 a).

—— A new interpretation of plasmodesmatal ultrastructure. Planta **82**, 100 (1968 b).

ROBERTSON, J. D.: The organization of cellular membranes. In: Molecular organiza- and biological function, p. 65. New York and London: Harper and Row (1967).

ROBERTSON, M. M., and R. C. KIRKWOOD: Differential uptake and movement as a factor influencing selective toxicity of MCPA and MCPB. 8th Brit. Weed Control Conf. Proc., p. 269 (1966).

ROELOFSEN, P. A.: The plant cell-wall. In: Handbuch der Pflanzenanatomie, vol. 3, p. 1. Berlin-Nikolassee: Gebrüder Borntraeger (1959).

ROGERS, H. J., and H. R. PERKINS: Cell walls and membranes. London: E. and F. N. Spon (1968).

ROHRBACH, P. W.: Penetration and accumulation of petroleum spray oils in the leaves, twigs, and fruit of citrus trees. Plant Physiol. **9**, 699 (1934).

Rossi, G., and O. Arrigoni: Ricerche sull'ultrastruttura della parete cellulare. G. Bot. Ital. **72**, 271 (1965).

Roux Lopez, J.: Estudio Morfologico de la Epidermis de Algunas Xerofitas Mexicanas. Thesis, Univ. Nac. Mex., Mexico City (1964).

Russell, R. S.: The extent and consequences of the uptake by plants of radioactive nuclides. Ann. Rev. Plant Physiol. **14**, 271 (1963).

Sabnis, D. D., and L. J. Audus: Growth substance interactions during uptake by mesocotyl segments of *Zea mays* L. Ann. Bot. **31**, 263 (1967 a).

—— —— Growth substance interactions during uptake by coleoptile segments of *Avena sativa* and hypocotyl segments of *Phaseolus radiatus*. Ann. Bot. **31**, 817 (1967 b).

Sacher, J. A.: The effect of free space enzymes on uptake of organic molecules. In: Absorption and translocation of organic substances in plants. 7th Ann. Symp., Amer. Soc. Plant Physiol. S. Sect., p. 29 (1964).

Sachs, R. M., Y. Shia, and R. G. Maire: Penetration, translocation, and metabolism of ^{14}C-Alar (B-9), a plant growth retardant. Plant Physiol. **42**, S-50 (1967).

Sargent, J. A.: The penetration of growth regulators into leaves. Ann. Rev. Plant Physiol. **16**, 1 (1965).

—— The physiology of entry of herbicides into plants in relation to formulation. 8th Brit. Weed Control Conf. Proc., p. 804 (1966).

——, and G. E. Blackman: Studies on foliar penetration. 2. The role of light in determining the penetration of 2,4-dichlorophenoxyacetic acid. J. Expt. Bot. **16**, 24 (1965).

Saunders, P. F., C. F. Jenner, and G. E. Blackman: The uptake of growth substances. IV. Influence of species and chemical structure on the pattern of uptake of substituted phenoxyacetic acids by stem tissues. J. Expt. Bot. **16**, 683 (1965 a).

—— —— —— The uptake of growth substances. V. Variation in the uptake of a series of chlorinated phenoxyacetic acids by stem tissues of *Gossypium hirsutum* and its relationship to differences in auxin activity. J. Expt. Bot. **16**, 697 (1965 b).

—— —— —— The uptake of growth substances. VI. A comparative study of the factors determining the patterns of uptake of phenoxyacetic acid and 2,4,5-trichlorophenoxyacetic acid, weak and strong auxins, by *Gossypium* tissues. J. Expt. Bot. **17**, 241 (1966).

Schlafke, E.: Kritische Untersuchungen zur wanderung von Fluorochromen in Blättern. Planta **50**, 388 (1958).

Schmid, W. E.: On the effects of DMSO in cation transport by excised barley roots. Amer. J. Bot. **55**, 757 (1968).

Schnepf, E.: Untersuchungen über Darstellung und Bau der Ektodesmen und ihre Beeinflussbarkeit durch stoffliche Faktoren. Planta **52**, 644 (1959).

—— Zur Feinstruktur der Drüsen von *Drosophyllum lusitanicum*. Planta **54**, 641 (1960).

—— Zur Cytologie und Physiologie pflanzlicher Drüsen. III. Cytologische Veränderungen in den Drüsen von *Drosophyllum* während der Verdauung. Planta **59**, 351 (1963).

—— Zur Feinstruktur der schleimsezernierenden Drüsenhaare auf der Ochrea von *Rumex* und *Rheum*. Planta **79**, 22 (1968).

Schroeter, C.: Das Pflanzenleben der Alpen. Eine Schilderung der Hochgebirgsflora. Zürich: Verlag Albert Raustein (1923).

Schumacher, W.: Untersuchungen über die Wanderung des Fluoresceins in den Harren von *Cucurbita pepo*. Jahrb. Wiss. Bot. **82**, 507 (1936).

—— Der Stofftransport zwischen parenchymatischen Zellen (Nahtransport). In: Handbuch der Pflanzenphysiologie. XIII. Der Stofftransport in der Pflanze, p. 3. Berlin, Heidelberg, New York: Springer-Verlag (1967 a).

—— Der Transport von Fluorescein in Haarzellen. In: Handbuch der Pflanzen-physiologie. XXII. Der Stofftransport in der Pflanze, p. 17. Berlin, Heidelberg, New York: Springer-Verlag (1967 b).

Sciuchetti, L. A.: The effects of DMSO alone and when combined with various growth regulators on the growth and metabolic products of Datura spp. In: Biological actions of dimethyl sulfoxide. Ann. N. Y. Acad. Sci. 141, 139 (1967).

——, and G. G. Hutchison: The influence of gibberellic acid and dimethylsulfoxide (DMSO) on the growth and metabolic products of Datura tatula. Lloydia 29, 368 (1966).

Scott, F. M.: Lipid deposition in intercellular space. Nature 203, 164 (1964).

—— Cell wall surface of the higher plants. Nature 210, 1015 (1966).

Setterfield, G., and S. T. Bayley: Structure and physiology of cell walls. Ann. Rev. Plant Physiol. 12, 35 (1961).

Sharma, G. K., and D. B. Dunn: Effect of environment on the cuticular features in Kalanchoe fedschenkoi. Bull. Torrey Bot. Club 95, 464 (1968).

Shellhorn, S. J., and H. M. Hull: A six-dye staining schedule for sections of mesquite and other desert plants. Stain Technol. 36, 69 (1961).

—— —— Enhanced herbicidal absorption in mesquite with a DMSO carrier complex. Weed Sci. Soc. Amer. Abstr. No. 163 (1969).

Siddiq, E. A., R. P. Puri, and V. P. Singh: Studies on growth and mutation frequency in rice in treatments with dimethyl sulfoxide and ethyl methane sulphonate. Current Sci. 37, 686 (1968).

Siegel, S. M.: The plant cell wall. A topical study of architecture, dynamics, comparative chemistry and technology in a biological system. New York: Macmillan (1962).

Sievers, A.: Über den Einfluss von monochromatischem Licht auf die Darstellbarkeit der Ektodesmen. A. Naturforsch. 15b, 49 (1960).

—— Zur Epidermisaussenwand der Fühlborsten von Dionaea muscipula. Planta 83, 49 (1968).

Sifton, H. B.: On the hairs and cuticle of Labrador tea leaves. A developmental study. Can. J. Bot. 41, 199 (1963).

Silva Fernandes, A. M. S.: Studies on plant cuticle. VIII. Surface waxes in relation to water-repellency. Ann. Applied Biol. 56, 297 (1965 a).

—— Studies on plant cuticle. IX. The permeability of isolated cuticular membranes. Ann. Applied Biol. 56, 305 (1965 b).

——, R. F. Batt, and J. T. Martin: The cuticular waxes of apple leaves and fruits and the cuticles of pear fruits during growth. Ann. Rept. Agr. Hort. Research Sta., Long Ashton, Bristol, p. 110 (1963).

Simon, E. W., and H. Beevers: The effect of pH on the biological activities of weak acids and bases. I. The most usual relationship between pH and activity. New Phytol. 51, 163 (1952).

Sirois, D. L.: Toxicity of surfactant-herbicide combinations to Lemna minor L. Ph.D. Diss., p. 97. Iowa State Univ. Sci. Technol. (1967).

Sitte, P.: Feinbau und Funktion der Pflanzenzellwand. Die Umsch. in Wiss. und Tec. 9, 273 (1962).

——, and R. Rennier: Untersuchungen en cuticularen Zellwandschichten. Planta Arch. Wiss. Bot. 60, 19 (1963).

Sjöstrand, F. S.: Critical evaluation of ultrastructural patterns with respect to fixation. In: Symp. Internat. Soc. Cell Biol., vol. I, p. 47. The interpretation of ultrastructure. New York: Academic Press (1962).

Skoss, J. D.: Structure and composition of plant cuticle in relation to environmental factors and permeability. Bot. Gaz. 117, 55 (1955).

Smith, G. N., B. S. Watson, and F. S. Fischer: Investigations on Dursban insecticide. Uptake and translocation of [^{36}Cl] O,O-diethyl O-3,5,6-trichloro-2-

pyridyl phosphorothioate and [¹⁴C] O,O-diethyl O-3,5,6-trichloro-2-pyridyl phosphorothioate by beans and corn. J. Agr. Food Chem. 15, 127 (1967).

SMITH, L. W., and C. L. FOY: Penetration and distribution studies in bean, cotton, and barley from foliar and root applications of Tween 20-C¹⁴ fatty acid and oxyethylene labeled. J. Agr. Food Chem. 14, 117 (1966).

—— —— Interactions of several paraquat-surfactant mixtures. Weeds 15, 67 (1967).

—— ——, and D. E. BAYER: Structure-activity relationships of alkylphenol ethylene oxide ether non-ionic surfactants and three water-soluble herbicides. Weed Research 6, 233 (1966).

—— —— —— Herbicidal enhancement by certain new biodegradable surfactants. Weeds 15, 87 (1967).

SMITH, R. C., and E. EPSTEIN: Ion absorption by shoot tissue: Technique and first findings with excised leaf tissue of corn. Plant Physiol. 39, 338 (1964).

SOROKIN, H. P., and K. V. THIMANN: The histological basis for inhibition of axillary buds in *Pisum sativum* and the effects of auxins and kinetin on xylem development. Protoplasma 59, 326 (1964).

SPANSWICK, R. M., and J. W. F. COSTERTON: Plasmodesmata in *Nitella translucens*: Structure and electrical resistance. J. Cell Sci. 2, 451 (1967).

STACE, C. A.: Cuticular studies as an aid to plant taxonomy. Bull. Brit. Mus. (Nat. Hist.) Bot. 4, 1 (1965).

STADELMANN, E. J., and J. WATTENDORFF: Plasmolysis and permeability of alpha-irradiated epidermal cells of *Allium cepa*. Protoplasma 62, 86 (1966).

STAEHELIN, L. A.: The interpretation of freeze-etched artificial and biological membranes. J. Ultrastruct. Research 22, 326 (1968).

STEIN, H.: Actinomycin D: Its inhibitory effect on the development of epidermal hairs on seedlings of *Sinapis alba* L. Israel J. Bot. 16, 124 (1967).

STEIN, W. D.: The movement of molecules across cell membranes. New York: Academic Press (1967).

STEINER, A., L. PRICE, K. MITRAKOS, and W. H. KLEIN: Red light effects on uptake of ¹⁴C and ³²P into etiolated corn leaf tissue during photomorphogenic leaf opening. Physiol. Plant. 21, 895 (1968).

STEINHUBEL, G.: Vplyv voskoviteho povlaku na tepelny rezim listov mahonie. Biol. (Bratislava) 22, 246 (1967).

STEWART, J. M., and E. A. C. FOLLETT: The electron microscopy of leaf surfaces preserved in peat. Can. J. Bot. 44, 421 (1966).

STROEV, S. S.: Vliyanie poverkhnostonoaktivnykh veshchestv na katalaznuyu aktivnost' Candida albicans. Tr. Leningrad Khim-Farm Inst. 18, 109 (1965).

STRUGGER, S.: Luminiszenzmikroskopische Analyse des Transpirationsstromes in Parenchymen. Biol. Zent. 50, 409 (1939).

SUKHORUKOV, K. T., and Y. M. PLOTNIKOVA: O strukture i funktsiyakh plazmodesm i ektodesm. Usp. Sovrem. Biol. 60, 299 (1965).

SZABO, S. S.: The hydrolysis of 2,4-D esters by bean and corn plants. Weeds 11, 292 (1963).

TALTON, J. H., JR., C. H. HENDERSHOTT, and R. H. BIGGS: Absorption and translocation of iodoacetic acid by foliage of two sweet orange varieties. Amer. Soc. Hort. Sci. Proc. 90, 117 (1967).

TEMPLE, R. E., and H. W. HILTON: The effect of surfactants on the water solubility of herbicides, and the foliar phytotoxicity of surfactants. Weeds 11, 297 (1963).

THOMPSON, N. P.: Regeneration of xylem and auxin transport in peanut leaves. Amer. J. Bot. 55, 731 (1968).

THOMSON, W. W., and L. L. LIU: Ultrastructural features of the salt gland of *Tamarix aphylla* L. Planta 73, 201 (1967).

TIETZ, H.: Der mit ³²P markierte Diäthylthionophosphorsäureester des β-Oxäthylthioäthyläthers (Wirkstoff des systemischen Insekticides, "Systox"), seine Aufnahme in die höhere Pflanze und sein Wanderungsvermögen. Höfchen-Briefe **7,** 1 (1954).

TING, I. P., and W. E. LOOMIS: Diffusion through stomates. Amer. J. Bot. **50,** 866 (1963).

TORII, K., and G. G. LATIES: Dual mechanisms of ion uptake in relation to vacuolation in corn roots. Plant Physiol. **41,** 863 (1966).

TRIP, P., and P. R. GORHAM: Translocation of sugar and tritiated water in squash plants. Plant Physiol. **43,** 1845 (1968).

TROSHIN, A. S.: Problems of cell permeability. Oxford: Pergamon Press (1966).

TROUGHTON, J. H., and D. M. HALL: Extracuticular wax and contact angle measurements on wheat *Triticum vulgare* L. Austral. J. Biol. Sci. **20,** 509 (1967).

TSCHIRLEY, F. H.: Research Report . . . Response of tropical and subtropical woody plants to chemical treatments. Rept. CR 13-67, Agr. Research Service, *U. S. Department of Agriculture* (1968).

TUKEY, H. B., JR.: Leaching of metabolites from above-ground plant parts and its implications. Bull. Torrey Bot. Club **93,** 385 (1966).

——, R. A. MECKLENBURG, and J. V. MORGAN: A mechanism for the leaching of metabolites from foliage. In: Proc. Symp. Use of Isotopes and Radiation in Soil-plant Nutrition Studies, p. 371. Internat. Atomic Energy Agency (1965).

TURKINA, M. V., and S. V. SOKOLOVA: Transport sakharozy cheerez kletochnye membrany provodyashchikh tkanei. Fiziol. Rast. **14,** 425 (1967).

TURRELL, F. M.: Citrus leaf stomata: Structure, composition, and pore size in relation to penetration of liquids. Bot. Gaz. **108,** 476 (1947).

UMOESSIEN, S. N., F. M. ASHTON, and C. L. FOY: Effects of hydrophilic-lipophilic balance (HLB) of nonionic surfactants on phytotoxicity of linuron and prometryne to carrots. Weed Soc. Amer. Abstr., p. 15 (1967).

VEEN, R. VAN DER, and G. MEIJER: Light and plant growth. New York: Macmillan (1960).

VENIS, M. A., and G. E. BLACKMAN: The uptake of growth substances. IX. Further studies of the mechanism of uptake of 2,3,6-trichlorobenzoic acid by *Avena* segments. J. Expt. Bot. **17,** 790 (1966).

VERNON, L. P., and E. SHAW: Photochemical activities of spinach chloroplasts following treatment with the detergent Triton X-100. Plant Physiol. **40,** 1269 (1965).

VIEITEZ, E., J. MÉNDEZ, C. MATO, and A. VÁZQUEZ: Effect of Tweens 80, 40 and 20 on the growth of Avena coleoptile sections. Physiol. Plant. **18,** 1143 (1965).

VLITOS, A. J., and H. G. CUTLER: Plant growth regulating activity of cuticular waxes of sugarcane. Plant Physiol. (Suppl.) **35,** vi (1960).

VOELLER, B. R., M. C. LEDBETTER, and K. R. PORTER: The plant cell. Aspects of its form and function. In: The cell, vol. VI, p. 245. New York and London: Academic Press (1964).

VOEVODIN, A. V., and S. V. ANDREEV: Absorption of 2,4-D by the leaves of wild plants. Dokl. Akad. Nauk. SSSR **134,** 211 (1960).

VOLK, R., and C. MCAULIFFE: Factors affecting the foliar absorption of N^{15} labeled urea by tobacco. Soil Sci. Soc. Amer. Proc. **18,** 308 (1954).

WAIN, R. L., and G. A. CARTER: Uptake, translocation and transformations by higher plants. In: Fungicides. An advanced treatise. Vol. I. Agricultural and industrial applications; environmental interactions, pp. 561-611. New York and London: Academic Press (1967).

WALDRON, J. D., D. S. GOWERS, A. C. CHIBNALL, and S. H. PIPER: Further observations on the paraffins and primary alcohols of plant waxes. Biochem. J. **78,** 435 (1961).

WARD, T. M., and R. P. UPCHURCH: Herbicide adsorption. Role of the amido group in adsorption mechanisms. J. Agr. Food Chem. **13,** 334 (1965).

WAX, L. M., and R. BEHRENS: Absorption and translocation of atrazine in quackgrass. Weeds 13, 107 (1965).

WEAVER, R. J., G. ALLEWELDT, and R. M. POOL: Absorption and translocation of gibberellic acid in the grapevine. Vitis Ber. Rebensforsch. 5, 446 (1966).

WEBSTER, D. H.: Entry of 2,4-dichlorophenoxyacetic acid into lambkill leaves at various 2,4-D/Tween 20 ratios. Weeds 10, 250 (1962).

——, and H. B. CURRIER: Heat-induced callose and lateral movement of assimilates from phloem. Can. J. Bot. 46, 1215 (1968).

WEISER, C. J., L. T. BLANEY, and P. LI: The question of boron and sugar translocation in plants. Physiol. Plant. 17, 589 (1964).

WELCH, R. M., and E. EPSTEIN: The dual mechanisms of alkali cation absorption by plant cells: Their parallel operation across the plasmalemma. Nat. Acad. Sci. Proc. 61, 447 (1968).

—— —— The plasmalemma: Seat of the type 2 mechanisms of ion absorption. Plant Physiol. 44, 301 (1969).

WESTWOOD, M. N., L. P. BATJER, and H. D. BILLINGSLEY: Effects of environment and chemical additives on absorption of dinitro-o-cresol by apple leaves. Amer. Soc. Hort. Sci. Proc. 76, 30 (1960).

WHITESELL, J. H., and N. P. THOMPSON: The uptake of Azodrin in corn. Plant Physiol. 43, S-35 (1968).

WILKINSON, R. E.: Vegetative response of saltcedar (Tamarix pentandra Pall.) to photoperiod. Plant Physiol. 41, 271 (1966 a).

—— Seasonal development of anatomical structures of saltcedar foliage. Bot. Gaz. 127, 231 (1966 b).

WILLIAMS, M. W., G. C. MARTIN, and E. A. STAHLY: The movement and fate of sorbitol-C^{14} in the apple tree and fruit. Amer. Soc. Hort. Sci. Proc. 90, 20 (1967).

WILSON, K.: The growth of plant cell walls. Internat. Rev. Cytol. 17, 1 (1964).

WITTWER, S. H.: Foliar absorption of plant nutrients. Adv. Frontiers Plant Sci. 8, 161 (1964).

—— Foliar application of nutrients for horticultural crops. Internat. Symp. Trop. and Subtrop. Hort. Proc., Indian Hort. Soc. and Internat. Hort. Soc. (In press) (1969).

——, M. J. BUKOVAC, W. H. JYUNG, Y. YAMADA, R. DE, H. P. RASMUSSEN, S. N. H. MARIAM, and S. KANNAN: Foliar absorption—Penetration of the cuticular membrane and nutrient uptake by isolated leaf cells. Qualit. Plant. 14, 105 (1967).

——, and F. G. TEUBNER: Foliar absorption of mineral nutrients. Ann. Rev. Plant. Physiol. 10, 13 (1959).

WOODBINE, M.: Biogenesis of fatty acids. In: Fatty acids: Their chemistry, properties, production, and uses, 2nd ed., part III, p. 1967. New York: Interscience (1964).

WOOLLEY, J. T.: Relative permeabilities of plastic films to water and carbon dioxide. Plant Physiol. 42, 641 (1967).

WORLEY, J. F.: Rotational streaming in fiber cells and its role in translocation. Plant Physiol. 43, 1648 (1968).

WORTMANN, G. B.: Elektronenmikroskopische Untersuchungen der Blattoberfläche und deren Veränderungen durch Pflanzenschutzmittel. Z. Pflanzenkr. Pflanzenpathol. Pflanzenschutzber. 72, 641 (1965).

YAMADA, Y., W. H. JYUNG, S. H. WITTWER, and M. J. BUKOVAC: The effects of urea on ion penetration through isolated cuticular membranes and ion uptake by leaf cells. Amer. Soc. Hort. Sci. Proc. 87, 429 (1965).

——, H. P. RASMUSSEN, M. J. BUKOVAC, and S. H. WITTWER: Binding sites for inorganic ions and urea on isolated cuticular membrane surfaces. Amer. J. Bot. 53, 170 (1966).

Yamaguchi, S.: Analysis of 2,4-D transport. Hilgardia 36, 349 (1965).

Zelitch, I.: Environmental and biochemical control of stomatal movement in leaves. Biol. Rev. 40, 463 (1965).

—— Control of leaf stomata—Their role in transpiration and photosynthesis. Amer. Sci. 55, 472 (1967).

—— stomatal control. Ann. Rev. Plant Physiol. 20, 329 (1969).

Zemskaya, V. A., and Y. V. Rakitin: Lokalizatsiya 2,4-D Vkletkakh list'ev kukuruzy i podsolnechnika. Fiziol. Rast. 14, 1011 (1967).

Ziegler, H., and U. Lüttge: Die Salzdrüsen von *Limonium vulgare*. II. Die Lokalisierung des Chlorids. Planta 74, 1 (1967).

——, and I. Vogt: The exudation of ^{14}C-Flurenol by guttation. Z. Pflanzenkr. Pflanzenpathol. Pflanzenschutzber. 4, 115 (1968).

Zöttl, P.: Vitalfärbestudien mit Methylrot. Protoplasma 51, 465 (1960).

Zweep, W. van der: Laboratoriumversuche über die Interaktion zwischen Ammoniumthiocyanate bzw. N^6-Benzyladenin und Amitrol. Z. Pflanzenkr. Pflanzenpathol. Pflanzenschutzber. 3, 123 (1965).

Subject Index

151